Connectionist Representations of Tonal Music

CONNECTIONIST REPRESENTATIONS OF TONAL MUSIC

Discovering Musical Patterns by Interpreting Artificial Neural Networks

Michael R. W. Dawson

AU PRESS

Published by AU Press, Athabasca University
1200, 10011 – 109 Street, Edmonton, AB T5J 3S8

ISBN 978-1-77199-220-6 (pbk.) 978-1-77199-221-3 (PDF) 978-1-77199-222-0 (epub)
doi: 10.15215/aupress/9781771992206.01

Cover and interior design by Sergiy Kozakov
Printed and bound in Canada by Friesens

Library and Archives Canada Cataloguing in Publication

Dawson, Michael Robert William, 1959-, author
Connectionist representations of tonal music : discovering musical patterns by interpreting artificial neural networks / Michael R.W. Dawson.

Includes bibliographical references and index.
Issued in print and electronic formats.

1. Music—Psychological aspects—Case studies. 2. Music—Physiological aspects—Case studies. 3. Music—Philosophy and aesthetics—Case studies. 4. Neural networks (Computer science)—Case studies. I. Title.

ML3830.D39 2018 781.1'1 C2017-907867-4
C2017-907868-2

We acknowledge the financial support of the Government of Canada through the Canada Book Fund (CBF) for our publishing activities and the assistance provided by the Government of Alberta through the Alberta Media Fund.

Contents

Chapter 9: Connectionist Reflections

Figures

Tables

Acknowledgements

I have long been interested in the foundations of cognitive science, in particular the relationship between classical cognitive science and connectionist cognitive science (Dawson, 1998, 2004, 2013). I believe that these two approaches are more similar than one might believe from reading the literature. This belief is supported by a research methodology in which networks are first trained on classical tasks, and then have their internal structure interpreted. In many cases, one can find very classical theories inside networks, in spite of typical claims that connectionism is quite distinct from the classical approach. Several years ago, we began to explore this methodology by training networks on very basic musical tasks (Dawson, 2009; Yaremchuk & Dawson, 2005, 2008). While this research provided additional support for the general research position—we pulled very formal theories out of these musical networks—it raised some interesting new issues. In particular, we discovered that the formal properties inside the musical networks were frequently quite different from the typical formal properties described in Western music theory. This book provides an extended investigation of these results, and of their implications.

The ideas presented in this book have flourished because of interactions and research collaborations with a number of undergraduate and graduate students, including Joshua Hathaway, Luke Kersten, John Hoffman, Brittany Koch-Hale, Vanessa Yaremchuk, and Brian Dupuis. I have also been encouraged by some supportive members of the Department of Music at the University of Alberta: Guillaume Tardif and Michael Frishkopf.

I dedicate this book to William Wallace Bruce Dawson, who gave me my first exposure to, and my lasting interest in, music, science, and computers.

CONNECTIONIST REPRESENTATIONS OF TONAL MUSIC

Overture: Alien Music

Steven Spielberg's 1977 movie *Close Encounters of the Third Kind* declares that we are not alone, and we should not be afraid. The film follows ordinary people after they experience a close encounter with an unidentified flying object. After this experience, the protagonist of the movie, Roy Neary, becomes obsessed with seeing the UFO again, as well as with a strange shape that has a deep meaning that he cannot quite fathom. Eventually he realizes that this shape represents Devils Tower, Wyoming, which is the location selected for first contact between humans and aliens.

A famous short musical signal composed by John Williams plays a key role in the film. Williams strove to write a signal that was long enough to be set apart from the simplest musical elements (chords or intervals) but not so long as to exist on its own as a melody. He decided that these goals required a theme that was only five notes in length, and composed about 350 different five-note permutations. Spielberg preferred one in particular, and it became one of the most famous musical themes in film history. At the climax of the film, it is performed on an ARP 2500 synthesizer located on a massive runway constructed atop Devils Tower. The music greets, and communicates with, the alien visitors.

A young, fresh-faced Philip Dodds was the ARP engineer who set up the system used in the movie. When the original actor who was to play it became sick, Spielberg saw Dodds working on the machine, liked his look, and cast him in the role of playing the synthesizer. As the scene unfolds, Dodds plays the musical signal faster and louder while the chief scientist (Lacombe, played by François Truffaut) strides out along the runway. Eventually the enormous mother ship arrives, hovers over the runway, and loudly echoes the notes that Dodds plays. This musical mimicking quickly erupts into an interstellar jam session of increasing tempo and complexity. Awestruck and wide eyed, Dodds exclaims, "What are we saying to each other?" This is a very deep question indeed.

The intended message in the film is that music is—literally—a universal language, one shared by all intelligent life forms. That the alien ship generates the same notes, and that it jams with Dodds's character in the same musical system, supports this optimistic view. To answer Dodds's question a scientist standing beside him says, "Seems they're trying to teach us a basic tonal vocabulary." Another immediately adds, "It's the first day of school, fellas."

However, other intriguing scenarios exist; there are alternative musical possibilities that Spielberg did not explore in his film. Does the fact that both the ARP 2500 and the mother ship generate the same basic tonal vocabulary imply that the humans and the aliens share the same underlying musical theory? Alternatively, is it possible that a completely different musical theory—an alien music theory that is dramatically different from our own—is still capable of generating the same patterns of musical notes? Perhaps all these fellas know they are attending the first day of school, but are not aware of the lecture topic!

Networks and New Musical Theory

The purpose of this book is to explore the possibility that different systems that are capable of generating the same musical inputs and outputs can do so by using quite different theories of music. Fortunately, instead of requiring a close encounter of the third kind for this exploration, this book adopts a more practical approach. A particular type of computer simulation, an artificial neural network, is taught to generate responses that are consistent with Western musical theory. For instance, the computer simulation is presented the tones that define a particular scale, and learns to respond with the tonic note of that scale, or to identify that scale as being major or minor.

However, when a network learns to generate the correct outputs to various musical inputs, it is not constrained by traditional Western music theory. Many researchers argue that the internal workings of artificial neural networks are quite distinct from the clear formal properties found in logic, mathematics, or music theory. As a result, it is possible that an artificial neural network can discover a completely different method —an alien or novel music theory—that generates the same input/output relationships as are defined by Western music theory.

In order to determine whether this is possible, it is necessary to examine the internal structure of a trained network to discover exactly how it generates musical responses. An artificial neural network is a messy collection of different processors (analogous to neurons) that send signals to one another through a larger and messier collection of weighted connections (analogous to synapses between neurons). The musical knowledge of a trained network lies in its internal patterns of connectivity. The messiness of this knowledge makes the existence of a new music theory possible. By making sense of a network's internal structure, we can discover the musical regularities that the network has learned, on its own, to exploit. Is the theory that the network learns the same as our own?

We will soon see that artificial neural networks can discover novel musical theories that seem quite different from typical accounts of Western music. This has interesting implications for music, insofar as it reveals alternative musical formalisms. This also has important implications for the study of musical cognition, because it reveals a variety of different kinds of representations that the human brain might use to process music.

Structure of the Book

The use of artificial neural networks to study human cognition is flourishing, and they are growing in importance in the study of musical cognition as well (Griffith & Todd, 1999; Todd & Loy, 1991). However, in this connectionist literature one rarely finds detailed interpretations of the internal structure of networks. A main purpose of this book is to provide numerous case studies of this approach, and to demonstrate its value to theories of music and to theories of musical cognition. I have hoped to convey this general message with the organization of the book.

The first two chapters provide a historical context for the current research. Chapter 1 provides a brief history of the scientific study of music and musical cognition, from the scientific revolution of the 17th century, through 19th century psychophysics, to modern musical cognitivism. One general theme that emerges from this discussion is that the psychology of music is intrinsically interdisciplinary; it involves the science of sound, the formal understanding of music theory, and the experimental understanding of music perception. It also makes clear that the modern psychology of music, musical cognition, views music perception as involving the active processing of auditory stimuli by organizing these stimuli using mental representations of music.

Chapter 2 then relates connectionist cognitive science, which uses artificial neural networks, to classical cognitive science, which views cognition as the rule-governed manipulation of symbols. Connectionism reacts against this symbolic view, and many musical connectionists seek to deal with regularities that cannot be captured formally. Chapter 2 introduces some basic properties of artificial neural networks, and introduces the methodology adopted in the chapters that follow. The chapter culminates in the argument that these networks should not be viewed as systems that are sensitive to informal properties of music. Instead, one should use these networks to inform both music theory and musical cognition by making sense of their internal structure. This requires a detailed understanding of how networks solve problems; this understanding will be formal. Networks do not merely capture informal properties of music; they capture new formal musical regularities.

The next six chapters provide case studies of this new approach to musical connectionism. Chapter 3 begins by introducing a case study in which a very simple network, called a perceptron, accomplishes a basic musical task: it learns to identify the tonic pitch of a presented scale. When we examine the connection weights of this network to attempt to explain how it works, we find a perfectly plausible formal musical interpretation of the perceptron's structure. However, this structure differs from the typical account of scale structure.

Chapter 4 continues with the study of scales by training a network to identify the mode of a presented scale, deciding whether a stimulus is a major scale or a harmonic minor scale. A more powerful artificial neural network, called a multilayer perceptron, is required to accomplish this task. The multilayer perceptron is more powerful than the perceptron discussed in Chapter 3 because it includes intermediate processors called hidden units. Interpreting this multilayer perceptron's internal structure reveals a quite novel account of the formal difference between major and minor keys, one that focuses upon the relationship between pairs of pitch-classes that are separated from one another by a particular musical interval, the tritone.

Chapter 5 provides another set of example networks that are trained on a particular task, called key-finding, which combines the tasks discussed in Chapters 3 and 4. To key-find a musical stimulus is to assert both the mode (major vs. minor) and the tonic of the scale used as the source of the musical stimulus's notes. The chapter begins with a multilayer perceptron that learns to generate the tonic and mode for a presented scale. It solves this problem with four hidden units that implement an interesting type of distributed representation called a coarse code. The interpretation of this network introduces the basic properties of coarse coding. Chapter 5 then considers a broader task: identifying the keys of various musical compositions. This is done using simpler networks, perceptrons, to create connectionist variants of three more traditional key-finding theories. The perceptrons identify the musical keys of a large number of test stimuli with a high degree of accuracy. The chapter ends by pointing out that an interpretation of the structure of these perceptrons suggests possible modifications to more traditional theories of key-finding.

Chapter 6 turns to the study of musical problems involving harmony: a combination of different pitches that occur simultaneously. It begins by introducing a very basic element of harmony, the triad, and defines four different triad types:

major, minor, diminished, and augmented. It then explores a multilayer perceptron that learns to identify these triad types. An examination of the structure of this network demonstrates that its hidden units apply the same "name" to a number of different pitch-classes that are related to one another by particular musical intervals. We call such an equivalence class a strange circle. The chapter then proceeds to define the properties of these strange circles and ends with another case study of a network trained to identify different types of chords, tetrachords. The analysis of this network's structure also reveals that it organizes inputs according to a variety of strange circles.

Chapter 7 provides a more complex example of a network trained to classify chords. It describes additional formulae for defining 12 different types of tetrachords for each of music's 12 major keys. It then reports the training of a multilayer perceptron that learned to classify an input tetrachord into these different tetrachord types. This more complex network requires seven hidden units to solve this classification problem. However, we can interpret its internal structure. This is because it also organizes input pitch-classes using strange circles. The interpretation of this network introduces an additional interpretative technique (examining bands in jittered density plots). As well, the structure of this extended tetrachord network provides another elegant example of coarse coding.

Chapter 8 explores an important source of tonality in Western music: musically related sequences of chords called chord progressions. The chapter begins by discussing an important chord sequence in jazz called the ii-V-I progression. We teach a number of different networks this progression; when provided one chord, a network responds with the next chord in the progression. A key difference between these various networks is that different codes are used to represent input and output chords. The question of interest is whether the choice of encoding affects the ease of learning the progression. Next, we turn to an elaboration of the ii-V-I progression and train networks on different encodings of a more complex jazz progression, the Coltrane changes. Again encoding is critical: the choice of encoding for the Coltrane changes determines not only the amount of learning but also the complexity of the network required to learn this progression. I will interpret a perceptron that learns one encoding of the Coltrane changes; this interpretation reveals once again the utility of the strange circles that we have already encountered in earlier chapters.

The final chapter of the book, Chapter 9, steps back to examine the general results reported in the various case studies. It begins with a discussion of the nature of this research program and then summarizes its most important results. It then turns to considering the implications of these results, first to the theory of music and then to the cognitive science of music. It concludes that our straightforward approach to musical networks has succeeded; even though the case studies involved training simple networks on basic properties of Western tonality, they led to the discovery a number of interesting, new musical properties. This success points the way to future musical studies; network interpretation is at the heart of each of these potential areas of investigation.

One purpose of this book is to present case studies that show that musical connectionism of the form championed here is viable. A second purpose is to inspire researchers to pursue similar studies, by studying the network architectures detailed in the chapters that follow or by exploring musical connectionism by interpreting different types of networks. All of the simulations that are described in this book were performed using networks whose properties have been described in detail elsewhere (Dawson, 2004). The software used to conduct these simulations has also been described in detail (Dawson, 2005) and is available free of charge from the author's website. Web resources for this book, which include links to software, the training files used to conduct the various simulations, resources for building new training files, and other relevant information are available here: http://www.bcp.psych.ualberta.ca/~mike/AlienMusic/

1

Science, Music, and Cognitivism

1.1 Mechanical Philosophy, Mathematics, and Music

Natural philosophy, developed by such giants as Copernicus, Galileo, Boyle, Newton, and Descartes, reigned during the scientific revolution from 1543 (the year of Copernicus's publication of *On the Revolutions of the Heavenly Spheres*) to 1687 (the year of Newton's publication of *Mathematical Principles of Natural Philosophy*) (Shapin, 1996). Shapin notes that the natural philosophy that emerged during the scientific revolution could also be called mechanical philosophy because it recognized that a variety of machines could be created that appeared to be purposeful, intentional, or sentient. Early natural philosophers were inspired by the properties of clocks because during the scientific revolution a variety of clockwork automata were created (Wood, 2002).

Viewing the world as a clock, natural philosophy embraced mathematics for describing and explaining nature's workings (Shapin, 1996). Mathematics played a key role in the theories of Galileo, Bacon, and Boyle; Newton's discoveries revealed that physical laws expressible in mathematical form govern the universe. Music played a key role in the pursuit of the mathematical explanation of nature. Kepler sought musical harmony in the motions of the planets (Stephenson, 1994). Lagrange used musical sound to link properties of his calculus to the physical world (Dhombres, 2002). As a student, Newton explored various mathematical means for dividing the octave into smaller musical intervals (Isacoff, 2001). Many natural philosophers believed that music provided evidence of the mathematical perfection of the natural world.

Of course the relation between mathematics and music originated long before natural philosophy. Around 500 BC, Pythagoras linked perceived pitch to the frequency at which a string vibrated. The Pythagoreans also determined that the most consonant musical intervals are ratios of string lengths that involved simple whole integers: 1/1 for unison, 2/1 for the octave, 3/2 for the perfect fifth, and 4/3 for the perfect fourth. In contrast, the ratio for a very dissonant interval, the tritone, is 45/32. Pythagorean geometry led to the discovery of irrational numbers, but irrational numbers are also found in music, and are related to dissonant musical intervals (Pesic, 2010).

Pythagorean notions of consonance led to tuning discrepancies (Donahue, 2005). If one starts at one pitch, and moves to a note that is seven Pythagorean octaves higher, one does not reach exactly the same note that is produced by moving 12 Pythagorean perfect fifths higher. The two final notes differ by the so-called Pythagorean comma (which equals 24 cents, where 100 cents = 1 semitone). Thus, one cannot have a perfect musical tuning system that includes perfect ratios for both octaves and fifths. This taunted those who believed in the mathematical perfection of music or nature. If consonant ratios are mathematically perfect, then why can they not be used to tune instruments whose notes spanned multiple octaves? At the start of the scientific revolution, many scholars, motivated by this problem, attempted to develop alternative approaches to tuning (Cohen, 1984).

1.2 Mechanical Philosophy and Tuning

1.2.1 Tempered Scales

The new approaches to tuning that emerged during the scientific revolution were motivated by several profound changes in music (Cohen, 1984). First, music became polyphonic: more than one voice or instrument simultaneously performed different parts. As a result, musical harmony became central to music, and the consonant combination of multiple musical parts required a proper tuning system. Second, new intervals (thirds and sixths) became accepted as being consonant. The English composer John Dunstable popularized these intervals in the early 15th century. However, Pythagorean tuning ignores these intervals. New approaches to tuning had to ensure the consonance of these new intervals. Third, fixed intonation instruments—instruments with notes tuned to specific pitches that cannot be altered during performance, like the modern piano—were more central to music. In addition, the notes of these instruments ranged over several octaves.

All of these developments created a need for a practical solution to the Pythagorean tuning discrepancy. In general, mechanical philosophers addressed this problem by developing new methods for dividing the octave into smaller intervals to define sets of available musical notes: they invented tempered scales. The primary goal of a tempered scale is to remove the Pythagorean comma (Donahue, 2005). This is accomplished by ensuring that, from some starting note with a frequency f, the note an octave higher has a frequency of 2f. In other words, the octave has primacy. Then, some other notes are added; these notes conform to the Pythagorean interval ratios. Finally, the remaining notes are included. These notes involve intervals whose ratios necessarily depart from the Pythagorean ideals. In other words, a tempered scale is a compromise: it ensures that some intervals conform to the Pythagorean ratios but deforms other intervals to eliminate the Pythagorean comma.

One example of a tempered scale is called just intonation; it was explored by mathematicians of the early 17th century (Barbour, 1972). Just intonation produces consonant harmonies involving fifths and thirds, but at the same time can produce very dissonant harmonies for other musical intervals. Another example of a tempered scale is called mean-tone temperament, which was invented by Pietro Aron in the early 16th century (Barbour, 1972). Mean-tone temperament distorts perfect fifths. This tuning favours thirds over fifths, which is the opposite of what is found in Pythagorean tuning (Donahue, 2005).

One problem with tempered scales is that they are defined with respect to a particular musical key (i.e., a particular starting frequency f). This means that the notes that define a scale in one musical key differ from those that define a scale in another. To perform the same piece in a different musical key (i.e., to musically transpose the composition), the instrument must be re-tuned. Another tempered scale, equal temperament, offered a solution to this problem.

Equal temperament, first mentioned by French philosopher, theologian, and mathematician Marin Mersenne in 1636, divides the octave into 12 equal segments, each representing the musical interval of a semitone. Equal temperament solves the problem of musical transposition because one can move a composition from key to key without having to re-tune an instrument. As a result, it became an ideal tuning for keyboard instruments (Isacoff, 2001, 2011).

However, equal temperament brought with it a new set of problems. By dividing the octave into 12 equal segments, it introduces irrational interval ratios. Consider some base pitch with frequency f. The pitch an octave higher has a frequency of 2f, or, to make an explicit link to the mathematics of equal temperament, a frequency of 212/12f. In equal temperament the tone that is a semitone higher than f will have the frequency 21/12f, the tone two semitones higher than f will have the frequency 22/12f, and so on. The appearance of irrational ratios—defined as 2 raised to some x/12 power—had two negative consequences. First, calculating the desired frequencies—a requirement for actually tuning an instrument to equal temperament—was difficult. A number of specialized tools and methods had to be invented in the 18th century to deal with this problem

(Scimemi, 2002). Equal temperament existed as a theoretical notion for a long time before it became practical to use it to tune instruments in the late 19th century. Second, the presence of irrational ratios meant that many musical intervals deviated enough from the Pythagorean ideals to sound less consonant. Indeed, the distorted ratios found in any tempered scale challenge the Pythagorean notion of consonance, as does the new musical aesthetic that considers intervals like thirds to be consonant.

This latter issue raises a theoretical problem concerning music that was also a concern of mechanical philosophers. Cohen (1984) calls it the problem of consonance. While the Pythagoreans identified particular frequency ratios as defining consonant musical intervals, they had no account of why this was so. The scientific revolution developed a very popular answer to this question that related consonance to the physical properties of sound. This answer is known as coincidence theory.

1.2.2 Coincidence Theory

In 1558, Gioseffo Zarlino explained the consonance of certain Pythagorean ratios with what is known as his scenario (Cohen, 1984). According to Zarlino, the consonant Pythagorean ratios all involve the first six integers; the set of integers from 1 to 6 defined the scenario. Zarlino proposed that any ratio of numbers that belonged to the scenario would be consonant, and supported this proposal with a variety of mystical arguments about why the members of the scenario were perfect numbers. In spite of this appeal to mysticism, coincidence theory was the most popular theory of consonance that emerged during this period (Cohen, 1984). It was then adopted, extended, and popularized by many leading figures of the scientific revolution including Galileo, Mersenne, Descartes, and Euler.

Coincidence theory attempts to establish the physical basis of consonance (Cohen, 1984). It begins with the observation that a plucked string produces sound by striking or percussing the surrounding air. We hear sound when these percussions reach our ears. Coincidence theory then recognizes that strings vibrating at different frequencies generate percussions at different rates. In some instances, the percussions generated by two different sound sources will reach the ears at the same time; these coincident percussions will be pleasing to the ear, or consonant. When the coincidence of percussions diminishes, so too will consonance. In other words, coincidence theory linked the purely mathematical view of consonance developed by the Pythagoreans to physical properties of sound vibrations, as they were understood in the 16th and 17th centuries.

1.3 Psychophysics of Music

The study of music during the scientific revolution produced new approaches to tuning and developed a theory that attempted to explain consonance in terms of coincident patterns of vibrations (Cohen, 1984). From this perspective, musical properties were physical properties of the world that could be studied scientifically and described mathematically. For instance, consonance was related to physical vibrations, and was not a construction of the mind.

1.3.1 Psychophysics

The systematic psychological study of consonance did not begin until the latter half of the 19th century; the golden age of this research occurred between 1840 and 1910 (Hui, 2013). It was during this period that psychophysical explorations of music began.

Psychophysics was invented by Gustav Fechner, who began by searching for relationships between physical properties of stimuli and the properties of the experiences that they produced (Fechner, 1966/1860). The psychophysical study of music began with attempts to identify universal mathematical laws that related physical properties of sound and music to mental properties of musical experience (Hui, 2013).

To Hui (2013), the psychophysical study of sound sensation is in turn related to an understanding of musical aesthetics. Her general argument is that the psychophysical study of music involved a constant tension between universal psychophysical laws and individual aesthetic responses, a conflict that could only be resolved by acknowledging that psychological processes contribute to the experience of music. Crucially, this meant that the physical properties of sound were not the only determinants of such phenomena as consonance.

At the same time, new theories of tuning became practical realities. When psychophysicists began to study music, equal temperament was rising in popularity (Isacoff, 2001). In short, psychophysicists studied music at a time when radically new notions of consonance and dissonance were emerging. How could one reconcile a natural science of music or consonance with the many changes in music and musical preference arising in the latter half of the 19th century?

1.3.2 On the Sensations of Tone

One of the most influential accounts of musical psychophysics (Hiebert, 2014) was Hermann Helmholtz's book *On the Sensations of Tone as a Physiological Basis for the Theory of Music* (Helmholtz & Ellis, 1863/1954). Helmholtz wove three different threads together. One concerns the physics of sound, musical sound in particular, and includes details about various devices for sound production and measurement. A second concerns the physiology of hearing, and includes a detailed account of the structure and function of the cochlea and the basilar membrane. A third concerns the implications of the properties of sound, and the physiology of hearing, for the perception of music, and provides detailed discussions of different tunings and musical aesthetics.

The core idea that Helmholtz uses is a reinvention of coincidence theory. Coincidence theory focused on the relationship between two pure tones. Helmholtz recognized that in almost every case a musical instrument will not generate a pure sine wave with frequency f. Sympathetic vibrations add to this fundamental frequency additional sine waves at various frequencies defined by the octave (e.g., 2f, 3f, 4f, and so on). These additional frequencies, called partials or harmonics, occur at weaker intensities than the fundamental, and tend to be weaker and weaker as the partial frequency becomes higher and higher. Helmholtz's theory of consonance emphasized harmonic interference.

When two musical sounds occur together, not only do their fundamental frequencies interact but their partials interact as well. For Helmholtz, interference patterns among all of the frequencies that are present cause the consonance of combined musical tones. Interference between partials would result in beats that produced an effect of roughness; consonant tones had little roughness; dissonant tones had more roughness because of more interference among partial frequencies. Importantly, these interference patterns affected the perception of music because they were detectable by the physiological mechanisms of hearing (Hui, 2013).

Helmholtz provided a detailed analysis of beat patterns related to a variety of musical intervals, and used this analysis to make fine-grained predictions about consonance. His quantitative account of consonance coincides with a variety of experimental studies of consonance conducted in the 20th century (Krumhansl, 1990a). Helmholtz used his new theory of consonance to inform musical aesthetics and to defend his personal views on the beautiful in music. Helmholtz was a champion of just intonation (Hiebert, 2014; Hui, 2013) and was highly critical of the rising popularity of equal temperament. He believed that just intonation produced music that was far more consonant than was possible in equal temperament. Hui points out that Helmholtz makes the physiological argument that just intonation produces music more consistent with the physiology of hearing than is produced using equal temperament. However, Hui also notes that Helmholtz recognized that there were individual differences in musical aesthetics. How was this to be reconciled with his detailed psychophysical theory?

Helmholtz argued that while his psychophysical account of the sensations of tone could inform the elementary rules of musical composition, these rules of composition are not natural laws. Different listeners were capable of enduring (and appreciating!) different degrees of roughness. Individual differences in musical tastes were a function of experience, culture, and education. One's musical knowledge, culture, or experience could influence the aesthetic experience of consonance and dissonance.

1.3.3 Individual Contributions

Helmholtz's influential studies suggest that musical perception is only based in part on universal laws involving the physics of sound and the physiology of hearing. Music perception is also affected by aesthetic considerations that stem from an individual's own culture, experience, and expertise. This position coincided with the tenets of musical Romanticism, which arose during the same era as musical psychophysics. Musical Romanticism emphasized individuality. That is, in Romanticism individual composers aimed to communicate their personal emotions and imaginations; Romanticism also heralded the virtuoso instrumentalist (Longyear, 1988). Informed by the theories of music critics like Eduard Hanslick (Hanslick, 1854/1957), audiences began to carefully listen to music—not just to enjoy it but also to understand it (Hui, 2013). Such intellectual listening depends heavily upon an individual's musical tastes and knowledge.

One consequence of psychophysicists accepting that an individual's mind or knowledge greatly affects musical perception is the need to discover the psychological laws that govern these influences. This research goal is central to the cognitive study of music that began in the middle of the 20th century. The cognitive approach recognized that there could exist abstract laws governing perception and thinking, but these laws are in turn linked to physical mechanisms. In the next section, we briefly introduce this cognitive approach and then describe the kinds of theories it produced to account for musical perception.

1.4 From Rationalism to Classical Cognitive Science

1.4.1 Rationalism

Mechanical philosophy was not only interested in explaining the natural world but also in explaining the mind, as exemplified in the 17th century philosophy of René

Descartes (Descartes, 1637/2006, 1641/1996). To establish a rigorous philosophy, Descartes adopted the mathematical approach of derivation from axioms. The basis of his philosophy was the axiom that he was a thinking thing: "A thing that doubts, understands, affirms, denies, is willing, is unwilling, and also imagines and has sensory perceptions" (p. 19). Cartesian philosophy proceeded by using this axiom to prove the existence of a perfect God who would not deceive, and then to establish the existence of an imperfectly sensed external world. As this philosophy derives new truths from axioms, we know it as rationalism. Cartesian philosophy has continued to inspire the study of mind centuries after its invention. In particular, it provides the philosophical foundations for modern cognitivism.

1.4.2 Cognitivism

For most of the first half of the 20th century, experimental psychology was dominated by an approach called behaviourism (Boring, 1950; Leahey, 1987). Behaviourism focused on investigating mechanistic links between observables—observable environmental stimuli and observable responses to these stimuli from organisms. Behaviourists strove to exclude elements that could not be directly observed (such as internal mental states) from their theories.

The 20th century invention of the digital computer made possible a new approach to studying psychology—cognitivism. Cognitivism reacted against behaviourism. In explaining the operations of a digital computer, one typically appeals to internal (and directly unobservable) states: the information represented inside the computer's memory (the symbols) and the operations used to manipulate this internal information (the rules). Cognitive psychology explored the hypothesis that thinking was identical to the operations of a digital computer. That is, cognitivism hypothesized that cognition is literally the rule-governed manipulation of mentally represented information.

Inspired by the digital computer, what is now known as classical cognitive science is a modern descendant of Cartesian philosophy (Dawson, 1998, 2013; Dawson, Dupuis, & Wilson, 2010a). In viewing thought as analogous to the rule-governed operations of a digital computer, classical cognitive science is a modern variant of viewing thinking as analogous to axiomatic derivations. In accordance with mechanistic philosophy's affinity for mathematics, and with Descartes's axiomatic philosophy, classical cognitive science often uses mathematics or logic to investigate cognitive phenomena (Dawson, 1998).

Nineteenth-century psychophysics is also a precursor to modern cognitivism. Helmholtz argued that hearing served to build internal representations of the world.

That is, the result of audition is to produce "mental images of determinate external objects, that is, in perceptions" (Helmholtz & Ellis, 1863/1954, p. 4). Behaviourism, in contrast, attempted to expunge mentalistic terms from its vocabulary (Watson, 1913). Behaviourism rejected using mentalistic terms like "unconscious inference" or "mental image"—terms that were central to Helmholtz.

1.5 Musical Cognitivism

1.5.1 Active Musical Processing

It has been argued that the study of music offers the strongest challenge against psychological behaviourism (Serafine, 1988). In Serafine's view, the creation and appreciation of new compositions, or of new musical styles, is unexplainable in terms of behavioural reinforcement. It instead demands a representational theory. For this reason, behaviourist accounts of the psychology of music are nonexistent. When behaviourism dominated psychology, musical psychologists were not as interested in explaining musical perception as they were in designing tests to measure individual differences in musical ability (Larson, 1930; Seashore, 1915, 1936, 1938/1967; Stanton & Seashore, 1935).

What is a fundamental difference between cognitive theories that appeal to mental representations and those, such as behaviourism, which do not? According to the psychology texts that emerged after the cognitive revolution in the latter half of the 20th century (Lindsay & Norman, 1972; Neisser, 1967; Reynolds & Flagg, 1977), the key difference is active processing. Cognitivists argued that behaviourism treated organisms as passive responders to the environment (Bertalanffy, 1967, 1969). In contrast, cognitivism assumed "a constantly *active* organism that searches, filters, selectively acts on, reorganizes and creates information" (Reynolds & Flagg, 1977, p. 11, their italics). For some, cognitivism was revolutionary not because it proposed representations but because it proposed active information processing.

According to classical cognitivism, organisms are active processors that first receive information from the environment, and then organize this information into representations of the world that can be used to think and plan, and finally perform some action on the world based upon this thinking and planning (Dawson, 2013). In other words, our experience of the world results from combining the information that we receive about the world with our existing beliefs, desires, and knowledge. Thus, modern cognitivism not only embraces the individual differences revealed by the psychophysical study of music (Hui, 2013), but also attempts to develop scientific theories of an individual's cognitive contributions to experience.

With the invention of new information-processing technology and new experimental methods, cognitivism went far beyond the psychophysical tradition of the 19th century. Modern cognitivists were not content to merely appeal to vague notions like unconscious inference. Instead, they were obligated to determine the nature of mental representations, as well as the kinds of rules used to manipulate them (Chomsky, 1965; Dutton & Starbuck, 1971; Feigenbaum & Feldman, 1995; Newell & Simon, 1972; Winston, 1975).

Unsurprisingly, the cognitive revolution has produced many important representational theories of music cognition (Cook, 1999; Deliège & Sloboda, 1997; Deutsch, 1982, 1999, 2013; Francès, 1988; Howell, Cross, & West, 1985; Krumhansl, 1990a; Lerdahl, 2001; Lerdahl & Jackendoff, 1983; Sloboda, 1985; Snyder, 2000; Temperley, 2001). In general, cognitive theories of music propose that musical perception involves organizing musical stimuli using existing mental representations. Musical cognitivists aim to explain the nature of musical representations, as well as the processes that manipulate these representations. The remainder of this section provides two examples to illustrate how classical cognitive scientists study music. We will see that these two topics will be important in later chapters when we begin to examine the musical regularities that artificial neural networks exploit.

1.5.2 The Tonal Hierarchy

Carol Krumhansl's investigations of the cognitive foundations of musical pitch provide a prototypical example of the representational approach to music cognition (Krumhansl, 1990a). Her research goal concisely defines classical cognitive science's perspective on music cognition: "to describe the human capacity for internalizing the structured sound materials of music by characterizing the nature of internal processes and representations" (Krumhansl, 1990a, p. 6).

To accomplish this goal, Krumhansl (1990a) makes a number of design decisions to choose from the vast possibilities available in terms of the musical stimuli to use, the musical responses to observe, and the choice of subjects to study, not to mention the description, analysis, and modelling of experimental results. After considering these possibilities, Krumhansl decides to explore cognitive representations of the pitch-classes used to define Western tonal-harmonic music.

Several factors guide this decision. First, even though musical pitch is related to a continuous physical property (sine wave frequency) and is therefore in principle infinitely variable, Western tonal music is experienced as involving only 12 different pitch-classes (Révész, 1954). This leads to the principle of octave equivalence (Forte,

1973; Patel, 2008; Roig-Francolí, 2008; Straus, 2005), which assigns pitches separated by multiples of an octave to the same pitch-class.

Second, even though the set of pitch-classes is very small, it serves as the foundation of Western tonal music. One can use combinations of pitch-classes to define complex musical objects. For instance, by using pairs of pitch-classes one can define musical intervals. By combining two or more musical intervals (e.g., by using three or four pitch-classes) one can define chords. A combination of seven pitch-classes defines a musical key that is built around a tonic note (one of the pitch-classes) and is associated with a particular sonority called its mode (which in Western music is typically either major or minor). In short, an understanding of the cognition of pitch should provide the foundation for an understanding of the cognition of more complex musical stimuli.

Third, while one constructs complex musical stimuli by combining pitch-classes together, there are powerful constraints on such combinations. Not every pitch-class combination in music is equally likely. In particular, consonant combinations are far more likely than dissonant combinations. Furthermore, establishing a particular tonality—a critical characteristic of Western music—requires that a subset of pitch-classes is more likely to be present in a composition, while at the same time the remaining pitch-classes are less likely to occur. In addition, if one only considers the to-be-included pitch-classes then some are more likely to occur than others are. In short, the cognition of Western tonal music attends to various relationships between pitch-classes, and one can explore these relationships by studying reactions to a very manageable stimulus set (i.e., the 12 pitch-classes).

Krumhansl's basic method for studying the cognition of musical pitch is called the probe tone method (Krumhansl & Shepard, 1979). The probe tone method involves a series of musical trials. On any given trial a musical context is established (e.g., by playing part of a musical scale, a chord, or some other musical stimulus). Then a single probe note—one of the pitch-classes—is played. A subject's task is to rate how well the probe note is related to the musical context. Typically, subjects make this rating on a seven-point scale where a rating of one indicates a very bad relationship, a rating of four indicates a moderate relationship, and a rating of seven indicates a very good relationship. Variations of this general paradigm permit subjects to make judgments about more complex musical stimuli (Krumhansl, 1990a).

The probe tone method reveals systematic relationships between pitch-classes within a particular musical context (e.g., a specific musical key). After establishing a musical key, the pitch-class given the highest rating of relationship to the context is the tonic of the key. For example, if the musical context establishes a key of A major, then the pitch-class A receives the highest rating, as is illustrated in Figure 1-1.

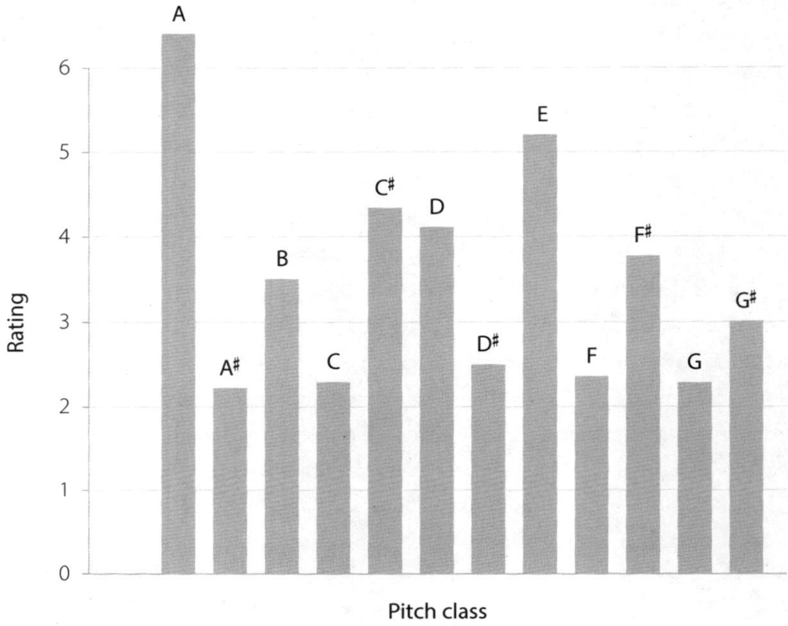

Figure 1-1 An illustration of "relatedness ratings" for each pitch-class in the context of the musical key A major.

The tones that receive the next highest ratings are those that belong in the third or fifth positions of the musical scale that defines the musical key. For the key of A major, these are the pitch-classes C♯ and E (see Figure 1-1). Receiving yet lower ratings are the remaining four pitch-classes that belong in the key's musical scale. For the key of A major, these are the pitch-classes B, D, F♯, and G♯. The pitch-classes not found in the key's scale elicit the lowest ratings. For the key of A major, these are the pitch-classes A♯, C, D♯, F, and G (see Figure 1-1).

The set of relationships between the 12 pitch-classes and a particular musical key is called the tonal hierarchy (Krumhansl, 1990a). If one changes the musical key, then one finds the same general pattern of ratings, but they are associated with different pitch-classes. For example, in the key of A major C♯ receives a very high rating, while C receives a very low rating (see Figure 1-1). However, in the key of C major C♯ receives the lowest rating while C receives the highest. In other words, each musical key is associated with its own tonal hierarchy (i.e., with a specific set of pitch-classes receiving high, moderate, low, and lowest ratings).

Krumhansl (1990a) has used the tonal hierarchies associated with different musical keys to explore higher-order relationships among musical keys. If two different keys are similar to one another, then their tonal hierarchies should be similar as well. In one study, the correlations between tonal hierarchies were calculated for every possible pair of the 12 different major and 12 different minor musical keys (Krumhansl & Kessler, 1982). Then a graphic representation of the similarity relationships between musical keys was determined by using multidimensional scaling to convert the correlations into a multidimensional map. In this map, a different point represents each key; points representing similar musical keys are nearer to one another in the map. Krumhansl and Kessler discovered that a four-dimensional map provided the best representation of the similarity data. This map arranged the keys in a spiral that wrapped itself around a toroidal surface.

The spiral arrangement of notes around the torus reflects elegant spatial relationships among tonic notes related to a standard device in music theory called the circle of fifths (Krumhansl, 1990a; Krumhansl & Kessler, 1982). The circle of fifths arranges the 12 pitch-classes around a circle; pitch-classes that are adjacent in the circle are a musical interval of a perfect fifth (seven semitones) apart. In the map discovered by Krumhansl and Kessler, the nearest neighbours for any key along the torus were its neighbouring keys in the circle of fifths. For instance, the nearest neighbours along the torus to the point representing the key of A major were the points for the keys of E major and D major. These two pitch-classes are on either side of A in the circle of fifths.

Krumhansl (1990a) summarizes a great deal of evidence in support of the tonal hierarchy as well as other organizational principles for music cognition. The tonal hierarchy is not a musical property per se, but is instead a psychologically imposed organization of musical elements. “The experience of music goes beyond registering the acoustic parameters of tone frequency, amplitude, duration, and timbre. Presumably, these are recoded, organized, and stored in memory in a form different from sensory codes” (Krumhansl, 1990, p. 281).

From Krumhansl’s (1990a) classical perspective, music cognition is a dynamic process in which mental representations relate musical sounds to one another and establish hierarchical relationships involving pitches, musical keys, intervals, and chords. The melodic, harmonic, and tonal roles of musical events are being continuously interpreted, organized, and structured. Importantly, these cognitive processes reflect the contributions of the individual to his or her own musical perceptions, contributions that were of great concern to psychophysical researchers (Hui, 2013) but which they did not explain in detail.

1.5.3 The Tritone Paradox

Krumhansl's (1990a) tonal hierarchy provides one prototypical example of how mental representations can explain musical cognition. The musical illusion called the tritone paradox provides another example. Below I will introduce the tones used to create this illusion, describe the illusion itself, and then provide a representational explanation of why the tritone paradox occurs.

The arrangement of the keys on a piano reveals two different properties. First, pitch ascends linearly from left to right along the keyboard. That is, after you play a note by pressing one of the piano keys, if you press the key immediately to its left, you will play a note that is a semitone lower; if you press the key immediately to its right, you will play a note that is a semitone higher. Second, pitch-class (the name of the note associated with any piano key) is arranged circularly. This is because the relationships between note names repeat themselves along the keyboard. For instance, find any key on the piano that plays an A. Its nearest neighbour to the left will be a key that plays a G♯; its nearest neighbour to the right will be a key that plays an A♯.

If you wished to use a single graph to depict these two relationships, then you would likely use a three-dimensional spiral like the one illustrated in Figure 1-2 (Deutsch, 2010). The vertical dimension of the spiral represents pitch. As you move upward from the bottom of this spiral toward the top, pitch gets higher. As a result, for any note in this spiral the note on one side will be higher in height (representing a pitch that is a semitone higher), while the note on the other side will be lower (representing a pitch that is a semitone lower).

In contrast, horizontal position around the spiral represents pitch-class relationships. No matter where you are in the spiral, one pitch-class will have the same neighbours. For instance, an A♯ will always have an A on its one side and a B on its other along the helix. Indeed, position in the horizontal dimension around the spiral is a reliable indicator of pitch-class. Imagine that the spiral in Figure 1-2 wraps itself around a cylinder. Every instance of one pitch-class is vertically aligned at the edge of this cylinder. For instance, the three different instances of A in Figure 1-3 are all stacked in the same column along the edge of an imaginary cylinder. In other words, position around the horizontal "circular dimension" of the Figure 1-2 spiral encodes pitch-class.

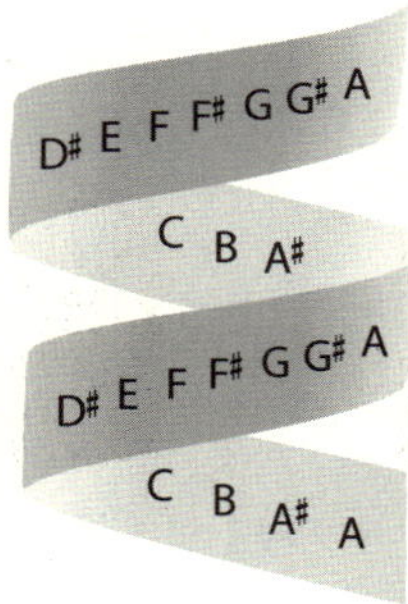

Figure 1-2 A three-dimensional spiral can simultaneously capture the linear arrangement of pitch and the circular arrangement of pitch-class on a piano keyboard.

Psychologist Roger Shepard reasoned that if he could eliminate the pitch dimension of music (e.g., if one were to compress the Figure 1-2 spiral to remove its height, so that all note names would fall around a single circle), then he could create an interesting musical illusion (Shepard, 1964). He used computer-generated sounds, now called Shepard tones, to test this hypothesis. A Shepard tone is a musical sound whose components all belong to the same pitch-class. For instance, a Shepard tone for the pitch-class A may be built from seven different pitches of A (e.g., A1, A2, A3, A4, A5, A6, and A7; where A4 is the A above middle C on the piano, A5 is the A that is an octave higher than A4, and A3 is the A that is an octave lower than A4). However, the loudness of each Shepard tone component decreases rapidly, moving in either direction away from the Shepard tone's central component. If one constructs a tone in the manner described above, the result is a sound that, when heard, specifies a particular pitch-class but does not clearly indicate the octave of the note being heard.

Shepard (1964) found that the Shepard tones produced a marked illusion of musical circularity. When Shepard presented 12 tones (one for each pitch-class) in sequence (e.g., A, A♯, B, C, D, D♯, E, F, F♯, G, and G♯), and then continued to repeat the sequence, listeners reported a definite linear progression of pitch. That is, subjects could hear successive notes increasing in pitch, with the current sound being a higher pitch than the one that preceded it. However, the experience puzzled the listeners. Some wondered why the pitch was increasing but did not really seem to get higher. "Some subjects were astonished to learn that the sequence was cyclic rather than monotonic and that in fact it repeatedly returned to precisely the tone with which it had begun" (Shepard, 1964, p. 2349). Shepard himself compared these auditory effects to the circular staircase illusions in the artwork of Maurits Escher.

Diana Deutsch used Shepard tones to discover another musical illusion called the tritone paradox (Deutsch, 1986, 1987, 1991). In this illusion, listeners hear pairs of Shepard tones and must judge whether the first or the second member of each pair has the higher pitch. In Deutsch's paradigm, the pitch-classes in each pair are separated by a musical interval called a tritone. If two pitches are a tritone apart, then they are six semitones apart, or separated by exactly half an octave. For example, A and D♯ are separated by a tritone, as are B and F.

Interestingly, the tritone paradox emerges by comparing the judgments of different subjects. For instance, one subject might judge that an A has a higher pitch than a D♯. However, a different subject might make exactly the opposite judgment from the same stimulus. "This demonstration is particularly striking when played to a group of professional musicians, who are quite certain of their own judgments and yet recognize that others are obtaining entirely different percepts" (Deutsch, 2010, p. 13).

How can one explain the tritone paradox? Deutsch (2010) proposes an elegant theory consistent with classical cognitive science. She suggests that pitch-class is represented by arranging pitch-classes around a circle that makes explicit the neighbourhood relations between pitch-classes. She further proposes that some notion of pitch height is also encoded in this representation. As a result, pitch-classes that fall on one side of this circular representation are judged higher in pitch than pitch-classes that fall on the other side of the circle.

This is illustrated in Figure 1-3. The top part of this figure arranges pitch-classes around a circle of minor seconds, so that neighbouring pitch-classes are a semitone apart. (This is identical to the neighbourhood relationship between note names found on a piano keyboard.) Pitch-classes joined by a diagonal through this circular arrangement are a tritone apart (e.g., C and F♯). In addition, a dashed line divides this representation into two, so that pitch-classes that fall above the dashed line are judged to have a higher pitch than pitch-classes that fall below the dashed line. The top part of Figure 1-3 illustrates a representation in which causes C to be judged to have a higher pitch than F♯.

However, Deutsch points out that there is no reason for there not to be individual differences in the orientation of this circle of pitch-classes. For instance, the bottom part of Figure 1-3 provides the same circular representation of pitch-classes, but it has been rotated 180° around the centre when compared to the top circle in the figure. As a result, different pitch-classes fall above the dashed line. A listener using the bottom representation of pitch-class would judge F♯ higher than C, which is opposite to the judgment obtained using the upper representation in the figure.

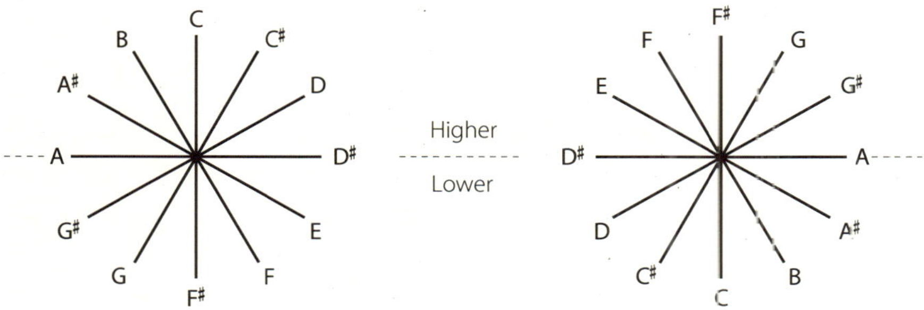

Figure 1-3 Using circles of minor seconds to explain the tritone paradox.

The tritone paradox, and Deutsch's explanation of this effect, provides a prototypical illustration of the classical approach to musical cognition. First, the physical properties of the musical stimulus are not sufficient to explain how it is experienced. For example, in the tritone paradox two different listeners can be presented exactly the same stimulus but have opposite experiences of it. As a result, an account of this phenomenon requires an appeal to internal processing of the external stimulus that produces the listener's final experience.

Second, the account of the tritone paradox depends on the properties of particular kinds of mental representations. When a listener hears a particular tone, the tone is mapped onto the particular location of a circle of pitch-classes like those illustrated in Figure 1-3. However, this circle of pitch-classes makes explicit other properties as well—in particular, the relative heights of different pitch-classes.

It is this additional information from the representation that results in the listener hearing a specific stimulus as having a lower pitch than the other has. Individual differences in a general property of this kind of representation (e.g., its orientation) explain individual differences in musical experiences. As was the case with Krumhansl's (1990a) tonal hierarchy, Deutch's explanation of the tritone paradox appeals to regularities governing representations but can also explain individual differences in these representations. Again, musical cognitivism's emphasis on active information processing embraces and explains the individual differences in perception that emerged from the psychophysical study of music (Hui, 2013).

1.6 Summary

The purpose of this chapter was to provide a historical overview of the science of music and to use this overview as a context for modern cognitivism. The chapter started with the perspective on music that began with the ancient Greeks, and which then flourished in the natural philosophy of the Enlightenment. According to this perspective, music was an attribute consistent with the perfect structure of nature, and all of its properties could be elegantly explained using mathematics. However, as natural philosophy matured, it made discoveries that challenged this view of music. In particular, the mathematical perfection of music was not reflected in various theories of consonance or of tuning. In the 19th century, other factors affected the account of music that mechanical philosophy held. In particular, changes in musical aesthetics as part of the rise of Romanticism, combined with discoveries of the new musical psychophysics, contradicted the view that music could be explained purely in terms of its mathematical or physical properties. In the latter half of the 19th century, it became clear that explanations of musical experience required appeals to the beliefs and cultures of individual listeners in addition to the physical properties of sound. Music was not merely physical: it was psychophysical.

The rise of cognitivism in the 20th century continued the tradition of explaining the perception and experience of music by appealing in part to psychological contributions of the listener. Classical cognitivism's approach to studying music represents a modern blending of the various traditions introduced in the current chapter. Like mechanical philosophy, musical cognitivism explains musical perception by appealing to formal or mathematical constructs. This is evident in its appeal to the rule-governed manipulation of mental representations. Like early musical psychophysics, musical cognitivism recognizes that perception of music depends on the interaction between the physical properties of acoustic stimuli and the organizational role of mental representations. This permits the study of musical cognition to appeal to scientific laws but also to be sensitive to individual differences. In short, musical cognitivism is an approach that is deeply rooted in natural philosophy but at the same time inspired by modern developments like the digital computer. In particular, musical cognitivism is primarily interested in identifying the nature of mental representations of music and the nature of the rules or operations that manipulate these representations in order to produce our musical experience.

However, the classical approach, inspired by the digital computer, is not the only school of thought in modern cognitivism (Dawson, 2013). Artificial neural networks form the basis of another, called connectionist cognitive science, which challenges some of the core assumptions of the classical approach. In addition, there is a growing interest in connectionist cognitive science in using artificial neural networks to capture some informal aspects of musical cognition, informal properties hypothesized to be beyond the reach of classical musical cognitivism. The purpose of this book is to explore some aspects of connectionist musical cognitivism; the next chapter provides an introduction to this general approach, as well as to the methodology employed in later chapters.

2

Artificial Neural Networks and Music

2.1 Some Connectionist Basics

2.1.1 Artificial Neural Networks

When classical cognitive science arose in the 1950s, it viewed cognition as being the rule-governed manipulation of symbols, analogous to the operations of a digital computer or to the definition of a formal logic. The view that thinking is performing a sort of mental logic is called logicism (Oaksford & Chater, 1991, 2007). Logicism, in the form of classical cognitive science, had many early and compelling successes, particularly in the computer simulation of higher-order cognition (Feigenbaum & Feldman, 1995; Newell & Simon, 1972).

However, while classical theorists promised that thinking machines were on the horizon, these promises were continually being broken. Some researchers began to question the foundations and the potential of classical cognitive science (Dreyfus, 1972, 1992). In particular, some challenged the notion that cognition is the rule-governed manipulation of symbols. Arguments arose that while the brain was likely an information processor, it was unlikely to be similar to a digital computer. As a result, some cognitive scientists adopted models of information processing that are more biologically plausible. These cognitive scientists are known as connectionists. In the mid-1980s, connectionist cognitive science arose as a reaction against its logicist ancestor.

Connectionist cognitive scientists employ artificial neural networks as models of human information processing (Dawson, 2004, 2005). An artificial neural network is a computer simulation of interconnected processing units. Each processing unit is analogous to a neuron and behaves as follows: First, it computes the total signal that it is receiving from other processors in the network. Second, the processor converts this total signal into some level of internal activity. Third, the processor unit sends its internal activity on to other processors. All of these operations are mathematical: signals and processor activities are all numbers that are determined by simple mathematical equations. In addition, these operations are parallel: many different processing units can be operating at the same time.

If processors in a PDP (Parallel Distributed Processing) network are analogous to neurons, then connections between processors are analogous to synapses between neurons. Each connection in a network has an associated weight (a numerical value) that indicates the connection's strength, as well as whether it is excitatory (positive weight) or inhibitory (negative weight). The connection is a communication channel that modifies a numerical signal sent through it by multiplying the signal by the connection's weight. In general, an artificial neural network has layers of processing units; signals pass through weighted connections from one layer to the next. The function of a typical network is to generate a desired response to a stimulus. The stimulus (e.g., a signal from the environment) is encoded as a pattern of activity in a layer of input units. The network's response to the stimulus is represented as a pattern of activity in its layer of output units. Intervening layers of processors in the system, called hidden units, detect more complex stimulus features.

Figure 2-1 provides an example of a musical artificial neural network. This network consists of 12 input units (the circles at the bottom of the figure), seven hidden units (the circles in the middle of the figure), and 12 output units (the circles at the top of the figure). Each line between circles in Figure 2-1 represents a weighted connection from one processor to another. In this network, each input unit has a connection to each hidden unit, and each hidden unit has a connection to each output unit. There are no direct connections between input and output units. This particular network is an example of a chord progression network that will be discussed later in this book. It is presented one chord from a musical sequence and responds with the next chord in the sequence. When a chord is presented to the input units (in this case by activating four pitches, B, C, E, and G, shown in grey in the figure), signals are sent through its layers producing responses in the output units. In Figure 2-1, the output units shaded in grey have turned on to the stimulus, while the unshaded output units remain off. Musically speaking, Figure 2-1 illustrates a network that has been presented a C major seventh (Cmaj7) chord and has responded with a C minor seventh (Cmin7) chord.

Figure 2-1 provides an example of one type of artificial neural network. There are many different types of networks, including distributed associative memories, feedforward networks, recurrent networks, and self-organizing maps (Amit, 1989; Anderson, 1995; Bechtel & Abrahamsen, 2002; Dawson, 2004; Gluck & Myers, 2001; Grossberg, 1988). Within each of these network types one finds many different variations, including different learning rules, numerous functions for computing incoming signals, various methods for computing processor activity, and so on. In other words, the domain "artificial neural network" is wide and varied. The current book explores a small subset of these possible network types in the context of music.

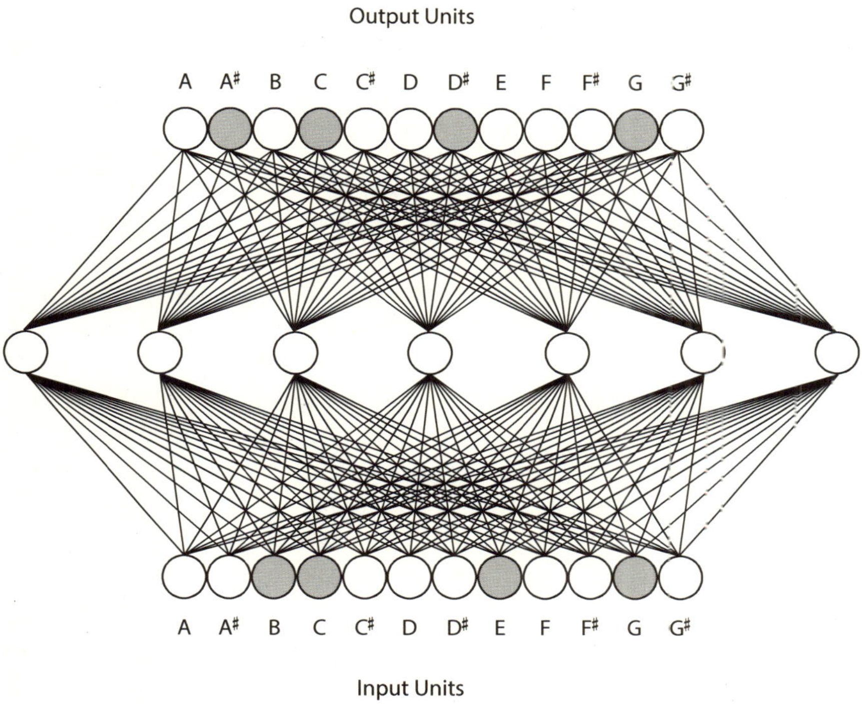

Figure 2-1 An example artificial neural network that, when presented a stimulus chord, responds with another chord.

2.1.2 Teaching Networks

How might a network like the one illustrated in Figure 2-1 "know" what chord to respond with when it is presented a stimulus? An artificial neural network's pattern of connectivity—its set of connection weights—defines its response to a stimulus. As a result, this pattern of connectivity is analogous to a computer program. However, one does not program artificial neural networks in any conventional sense. Instead, one teaches them. Networks receive a sequence of input patterns, and learn, by adjusting their connection weights, to produce the correct responses to presented patterns.

Typically, when a network is trained to classify patterns, it is presented a set of input patterns for which desired responses are already known. That is, each stimulus is associated with a desired response, and the goal is to train a network so that it generates this desired response for each pattern in the training set. To accomplish this, a supervised learning algorithm is used to train the network.

Supervised learning in general proceeds as follows: We start with a network whose connection weights are given small, random initial values. We present one of the training patterns to the network; it generates a response to this pattern using its current connection weights. Early in learning, because the network is started randomly, we expect its responses to be highly inaccurate. We can measure this inaccuracy by comparing the desired response for each output unit (the response that we want) to the observed response (the actual response generated by the network to the stimulus). We do so by taking the mathematical difference between the desired and observed responses. This difference is the error of an output unit.

Once error has been computed, it is used to modify connection weights in order to reduce network error. That is, after changing connection weights, the next time the same pattern is presented to the network the network will generate less error in response to it. There is a variety of different learning rules that can be used to train artificial neural networks (Bishop, 1995; Caudill & Butler, 1992; Grossberg, 1988; Ripley, 1996; Rojas, 1996; Shepherd, 1997). For networks that include hidden units, the error computed for each output unit must be sent backward through the network in order for hidden unit error to be determined, and for hidden unit connection weights to be modified (Rumelhart, Hinton, & Williams, 1986; Rumelhart & McClelland, 1986b). For all of the various supervised learning rules, there is one common feature: each time connection weights are modified network errors decrease. The goal is to reduce network error, with enough training, to a magnitude that is small enough to say that the network has learned the correct response to each of the training stimuli.

2.1.3 What Can Networks Do?

A connectionist network is a computer simulation that converts an input pattern into an output response. What kind of stimulus-response mappings can artificial neural networks learn to generate?

One common task is pattern recognition (Lippmann, 1989; Pao, 1989; Ripley, 1996). When a network performs pattern recognition, it identifies its input pattern as belonging to a particular class. For instance, one might present a network a set of musical notes that the network could then classify as representing a particular type of musical chord (Yaremchuk & Dawson, 2005). The Figure 2-1 network can be considered a pattern recognition system, because it generates a discrete or categorical response to a stimulus.

Another task that connectionist networks can accomplish is function approximation (Siegelmann, 1999; Takane, Oshima-Takane, & Shultz, 1994). In a function approximation task, a network maps an input pattern into a continuous output response. In general, the input units represent the values of one or more x-variables, and the output unit(s) represents some function of these variables. That is, the network computes a function y = f(x1, x2, . . . xn) where an output unit represents the value of y. For instance, later we will discuss a network that is presented a summary of the notes in a particular song, and generates the probability that the song is written in a particular key. Pattern recognition and function approximation are two very general tasks that artificial neural networks can accomplish very well. As a result, connectionist models have arisen in a variety of different research domains, including perception (Carpenter & Grossberg, 1992; Wechsler, 1992), language (Mammone, 1993), brain function (Amit, 1989; Burnod, 1990; Gluck & Myers, 2001), and animal learning (Dawson, 2008; Enquist & Ghirlanda, 2005; Schmajuk, 1997).

Connectionist models have also been applied to a wide variety of problems in music and in musical cognition (Bharucha, 1999; Fiske, 2004; Griffith & Todd, 1999; Todd & Loy, 1991). A variety of network architectures have been applied to such topics as classifying pitch and tonality, assigning rhythm and metre, classifying and completing melodic structure, and composing new musical pieces. Let us briefly consider some examples of musical connectionism.

Connectionist networks can accomplish a variety of tasks that require classification of basic elements of Western music (e.g., pitch, tonality, and harmony). Artificial neural networks have been trained to classify chords (Laden & Keefe, 1989; Yaremchuk & Dawson, 2005, 2008), to assign notes to structures similar to the tonal hierarchy (Leman, 1991; Scarborough, Miller, & Jones, 1989), to model the effects of musical expectations on musical perception (Bharucha, 1987; Bharucha & Todd, 1989), to add harmony to melodies (Berkeley & Raine, 2011; Shibata, 1991), to determine the musical key of a melody (Griffith, 1995), to identify a melody even when it has been transposed into a different key (Benuskova, 1995; Bharucha & Todd, 1989; Page, 1994; Stevens & Latimer, 1992), and to detect the chord patterns in a composition (Gjerdingen, 1992).

Artificial neural networks can also model some perceptual illusions involving pitch. One example is virtual pitch (Terhardt, Stoll, & Seewann, 1982a, 1982b). In this illusion, one constructs a musical signal from a combination of sine waves (i.e., harmonics) but does not include the lowest-frequency sine wave, the fundamental frequency. The fundamental frequency determines the pitch of the tone (i.e., the octave in which the pitch is experienced). Human listeners, however, do not hear a tone missing its fundamental frequency in a different octave. Instead, they hear the tone in the correct octave, as if the missing fundamental frequency is put back into the stimulus (Fletcher, 1924). Certain types of artificial neural networks can use context (i.e., the presence of harmonic sine waves) to add the missing fundamental (Benuskova, 1994; Sano & Jenkins, 1989).

Artificial neural networks can also handle other important aspects of music that are independent of tonality, such as assigning rhythm and metre (Desain & Honing, 1989; Griffith & Todd, 1999). For example, one network for assigning rhythm and metre uses a system of oscillating processors—units that fire at a set frequency (Large & Kolen, 1994). The phase of an oscillator's frequency can vary, and signals between processors enable their phases to entrain. This permits the network to represent the metrical structure of a musical input, even if the actual input is noisy or imperfect. This notion can be elaborated in a self-organizing network that permits preferences for, or expectancies of, certain rhythmic patterns to determine the final representation that the network converges to (Gasser, Eck, & Port, 1999).

The examples cited above generally involve using artificial neural networks to detect properties of existing music. The ability of networks to process tonality, harmony, metre, and rhythm also permits them to generate new music. Composition has in fact been one of the most successful applications of musical connectionism. Networks can compose single-voiced melodies on the basis of learned musical structure (Mozer, 1991; Todd, 1989); can compose harmonized melodies or multiple-voiced pieces (Adiloglu & Alpaslan, 2007; Bellgard & Tsang, 1994; Hoover & Stanley, 2009; Mozer, 1994); can improvise when presented new jazz melodies and harmonies (Franklin, 2006); and can improvise by composing variations on learned melodies (Nagashima & Kawashima, 1997).

The logic of network composition is that the relationship between successive notes in a melody, or between different notes played at the same time in a harmonized or multiple-voiced piece, is not random, but is instead constrained by stylistic, melodic, and acoustic constraints (Huron, 2006; Kohonen, Laine, Tiits, & Torkkola, 1991; Lewis, 1991; Mozer, 1991, 1994; Temperley, 2007). Networks can learn these constraints and then use them to generate the next note in a new composition.

The ability of artificial neural networks to exploit similarity relationships positions them to capture regularities that are difficult to express in language or using formal rules (Loy, 1991). This permits networks to solve musical problems that involve very abstract properties. For example, human subjects can accurately classify the genre or style of a short musical selection within a quarter of a second (Gjerdingen & Perrott, 2008). The notion of style or genre is too vague to be formalized in a fashion suitable for a classical rule-governed system (Loy, 1991). However, neural networks are up to the task, and can: classify musical patterns as belonging to the early works of Mozart (Gjerdingen, 1990); classify selections as belonging to different genres of Western music (Mostafa & Billor, 2009); evaluate the affective aesthetics of a melody (Cangelosi, 2010; Coutinho & Cangelosi, 2009; Katz, 1995); and even predict the possibility that a particular song has "hit potential" (Monterola, Abundo, Tugaff, & Venturina, 2009).

Artificial neural networks have musical applications that extend beyond human cognition. For instance, with the wide availability of digital music, networks are proving to be useful in serving as adaptive systems for selecting music, or generating musical playlists, based on a user's mood or past preferences combined with the ability to process properties of the stored music (Bugatti, Flammini, & Migliorati, 2002; Jun, Rho, & Hwang, 2010; Liu, Hsieh, & Tsai, 2010; Munoz-Exposito, Garcia-Galan, Ruiz-Reyes, & Vera-Candeas, 2007; Wieczorkowska & Kubera, 2010). Networks can also be used to automatically transcribe music (Marolt, 2004a, 2004b) and to generate realistic-sounding singing voices by manipulating vibrato (Gu & Lin, 2014).

Clearly, there is a great deal of interest in using artificial neural networks to study musical cognition. Bharucha (1999) provides five different advantages of connectionist research on music. First, artificial neural networks can account for how music is learned. Second, connectionist theories of such learning are biologically plausible. Third, networks provide accounts of music perception phenomena, such as contextual effects and the filling-in of incomplete information. Fourth, networks exploit similarity-based regularities that are important in theories of musical cognition. Fifth, networks may discover regularities (e.g., in musical styles) that elude more formal analyses.

This fifth observation made by Bharucha (1999) hearkens back to the tension, as discussed in Chapter 1, between universal laws and musical aesthetics faced by psychophysical researchers. Proposing that some aspects of music cannot be captured by formal rules is similar to claiming, like Helmholtz, that natural laws cannot explain musical aesthetics.

Much of the musical connectionism pursued in later chapters of this book reacts against this fifth point of Bharucha (1999). The current research does not agree that a main goal of musical connectionism is to capture informal regularities. Instead, the current research uses musical networks to reveal formal properties of music. The remainder of this chapter explores the uneasy relationship between formal and informal accounts of music, with a particular interest in connectionism's role in this relationship.

2.2 Romanticism and Connectionism

2.2.1 Musical Romanticism

At the end of the period stretching from 1543 to 1687, the scientific revolution evolved into the Enlightenment (Ede & Cormack, 2004). The Enlightenment saw many of the ideas born during the scientific revolution extended and modified, particularly with respect to individualism, freedom, politics, and commerce. The resulting Industrial Revolution transferred power and wealth from the nobility to the commercial class (Plantinga, 1984).

The ideas that characterize the Enlightenment profoundly influenced politics, thinking, and art, which, in turn, created discontentment with the existing social order, and led to the 1789 revolution in France. This, in turn, led to an artistic and intellectual movement called Romanticism (Claudon, 1980) that roughly spanned the period from just before the French Revolution through to the end of the 19th century. Because the Enlightenment evolved from the scientific revolution, it too exalted reason and rationality. In contrast, Romanticism emphasized the individual, the irrational, and the imaginative (Einstein, 1947; Plantinga, 1984). It replaced reason with an emphasis on the imaginary and the sublime. The Romantic artists looked back longingly at unspoiled, pre-industrial existence by depicting wild or fanciful settings. Nature was their inspiration. Romanticism appealed to the untamed mountains and chasms of the Alps to oppose the Enlightenment's view of an ordered, structured world.

Many argue that the most purely Romanticist art was music, because music expresses mystical or imaginative ideas and emotions that language cannot (Einstein, 1947; Plantinga, 1984; Sullivan, 1927). Language, of course, is a key vehicle of reason and rationality. In rejecting language, Romanticism focused upon purely instrumental music that "became the choicest means of saying what could not be said, of expressing something deeper than the word had been able to express" (Einstein, 1947, p. 32). Romanticist composers strove to replace the calculated, rational form of such music as Bach's contrapuntal fugues (Gaines, 2005; Hofstadter, 1979) with a music that expressed intense emotion and communicated the sublime (Einstein, 1947; Longyear, 1988; Plantinga, 1984; Whittall, 1987).

2.2.2 *Connectionism as Romanticism*

In reacting against classical cognitive science, connectionism also rejects the Cartesian rationalism that permeates classical cognitive science (Dawson, 2013). The rise of connectionist cognitive science is analogous to the Romanticist reaction against the Enlightenment. Dawson (2013) outlines many intellectual parallels between Romanticist music and connectionist cognitive science. We explore two of these parallels below.

The first parallel is the rejection of the logical. Romanticism rejected reason by moving away from language, particularly in music. This is paralleled in connectionism's claim that cognitive explanations need not appeal to explicit rules or symbols (Bechtel, 1994; Bechtel & Abrahamsen, 2002; Horgan & Tienson, 1996; Ramsey, Stich, & Rumelhart, 1991; Rumelhart & McClelland, 1986a). Connectionism abandons logicism, and assumes that the internal workings of its networks do not involve the rule-governed manipulation of symbols. "We would all like to attain a better understanding of the internal operations of networks, but focusing our search on functional equivalents to symbolic operations could keep us from noticing what is most worth seeing" (Bechtel, 1994, p. 458).

The second parallel is connectionism's sympathy with Romanticism's emphasis on nature. Cartesian philosophy, and the classical cognitive science that it inspired, view the mind as disembodied from the natural world. Connectionists reject this perspective by developing models that are biologically plausible or neuronally inspired (McClelland & Rumelhart, 1986; Rumelhart & McClelland, 1986b). Connectionism emphasizes the brain.

Biological inspiration carries with it sympathy with the sublime. Connectionists accept that the internal structure of their networks is very difficult to understand (Dawson, 1998, 2004, 2009; Dawson & Shamanski, 1994; McCloskey, 1991; Mozer & Smolensky, 1989; Seidenberg, 1993). This is because their networks mimic the mysterious structure of the brain. "One thing that connectionist networks have in common with brains is that if you open them up and peer inside, all you can see is a big pile of goo" (Mozer & Smolensky, 1989, p. 3). In the connectionist literature, detailed analyses of the internal structure of a network, coupled with accounts of how network structures solve problems of interest, are rare (Dawson, 1998, 2004, 2005, 2009, 2013; Dawson & Shamanski, 1994).

Connectionism's rejection of logicism, and its embrace of the sublime, accounts for its current popularity in the study of musical cognition. Some researchers believe that artificial neural networks can capture musical regularities that cannot be rationally expressed (Bharucha, 1999; Rowe, 2001; Todd & Loy, 1991). Of course, this belief about the utility of networks parallels the Romanticist view that one cannot formalize important characteristics of music.

This Romanticist perspective is readily evident, for example, in discussions of networks that compose music. Such networks are presumed to internalize constraints that are difficult to formalize. "Nonconnectionist algorithmic approaches in the computer arts have often met with the difficulty that 'laws' of art are characteristically fuzzy and ill-suited for algorithmic description" (Lewis, 1991, p. 212). Furthermore, these "laws" are unlikely to arise from analyzing the internal structure of a network, "since the hidden units typically compute some complicated, often uninterpretable function of their inputs" (Todd, 1989, p. 31). Such accounts of modern connectionist networks evoke the earlier musings of Helmholtz about the nature of musical aesthetics.

Connectionist Romanticism raises some questions that we explore in detail in the final section of this chapter. First, if the musical regularities captured by artificial neural networks cannot be formally expressed, then what is the purpose of such networks in a cognitive science of music? Second, is it possible that musical networks can capture formal musical regularities?

2.3 Against Connectionist Romanticism

2.3.1 Bonini's Paradox

Models attempt to enhance our understanding of the world. Cognitive scientists use many different kinds of models. These include statistical models that describe data (Kruschke, 2011; Lunneborg, 1994), mathematical models that provide quantifiable laws (Atkinson, Bower, & Crothers, 1965; Coombs, Dawes, & Tversky, 1970; Restle, 1971), and computer simulations that themselves generate behaviour of interest (Dutton & Starbuck, 1971; Newell & Simon, 1961, 1972). Regardless of type, a model serves to increase understanding by providing a simplified and tractable account of some phenomenon of interest.

Merely creating a model, however, does not always guarantee greater understanding. This is particularly true of computer simulations of cognitive processes (Lewandowsky, 1993). Such simulations can encounter what Dutton and Starbuck (1971) call Bonini's paradox. This paradox occurs when a computer simulation is as difficult to understand as the phenomenon being modelled. There are reasons to believe that the Romanticism of connectionism leads directly to Bonini's paradox, particularly in the study of musical cognition.

To begin, the internal structure of artificial neural networks is notoriously difficult to understand. This is because of their parallel, distributed, and nonlinear nature. Connectionists, after training a network, are often hard pressed to describe how it actually accomplishes its task.

In the early stages of the connectionist revolution, this was not a keen concern. The 1980s was a period of "gee whiz" connectionism (Dawson, 2009) in which connectionists modelled phenomena that were prototypical for classical cognitive science. In the mid-1980s, it was sufficiently interesting to show that such phenomena might be accounted for by alternative kinds of models. Researchers during this period were not required to delve into the details of the internal structures of networks to explain their operations. However, in modern connectionist cognitive science it is necessary for researchers to spell out exactly how networks function (Dawson, 2004). In the absence of such details, connectionist models have absolutely nothing to contribute to cognitive science (McCloskey, 1991). It is no longer enough for a network to be inspired by the (sublime) brain. A network must provide details that give insight into how brains might actually process information. An un-interpreted network produces Bonini's paradox.

Importantly, many musical networks have an additional wrinkle that makes them difficult to understand. In connectionism, there are two general approaches to network training. One is supervised learning. In supervised learning, a researcher defines a set of desired input/output pairings, and trains a network to generate these, typically with an error-correcting learning rule such as the one introduced in the next chapter. That is, in supervised learning the researcher knows beforehand what a network is designed to do and teaches the network to respond accordingly.

The other approach to network training is called unsupervised learning and is typically employed in what are called self-organizing networks (Amit, 1989; Carpenter & Grossberg, 1992; Grossberg, 1980, 1987, 1988; Kohonen, 1977, 1984). In unsupervised learning, one presents input patterns to a network, but these patterns are not paired with desired outputs. Instead, networks stabilize after every input and then modify their weights to encode this stable state. As a result, a self-organizing network learns the statistical regularities in its input patterns and generates responses that reflect these regularities, without external guidance or teaching.

In general, connectionist cognitive science is much more likely to use supervised learning than unsupervised learning. However, it has been argued that this is not true in the study of musical cognition (Dawson, 2013), which seems to have a marked preference for unsupervised learning. In two collections of papers about connectionist musical cognition (Griffith & Todd, 1999; Todd & Loy, 1991), one finds many more self-organizing networks than one would expect to find in other domains of cognitive science.

How is the use of self-organizing networks related to Bonini's paradox? First, one constructs these networks from the same components used to create other sorts of artificial neural networks. Therefore, they are just as difficult to interpret as any other kind of network. Second, if one uses unsupervised learning to train a network, then difficulties in understanding the network's internal structure are compounded by the additional fact that one may not even know what it is that the network has learned. If one does not know what a network's responses are supposed to signify, then how can one understand the network?

2.3.2 An Alternative Paradigm

Of course, a strong motivator for applying connectionism to musical cognition, and for preferring unsupervised learning, is the Romanticist view that important aspects of music are informal. Presumably, networks can capture these regularities informally. However, any input/output relationship that can be realized in an artificial neural network must be formal. All connectionist networks are mathematical engines that compute functions–they map numerical elements from an input domain onto numerical elements in an output domain. In other words, musical networks do not have the advantage of capturing informal regularities that symbolic languages cannot capture, as some propose (Bharucha, 1999). Instead, networks have the disadvantage of capturing formal regularities that are difficult to ascertain or to express. We must discover and detail these regularities if connectionism is to contribute to the cognition of music.

Certainly, networks are hard to interpret. However, it is not impossible to explore the internal structure of a trained network in order to explain how it converts its inputs into its responses. Connectionist cognitive scientists have developed many techniques for interpreting the internal structure of artificial neural networks (Baesens, Setiono, Mues, & Vanthienen, 2003; Berkeley, Dawson, Medler, Schopflocher, & Hornsby, 1995; Dawson, 2004, 2005; Gallant, 1993; Hanson & Burr, 1990; Hayashi, Setiono, & Yoshida, 2000; Hinton, 1986; Moorhead, Haig, & Clement, 1989; Omlin & Giles, 1996; Setiono, Baesens, & Mues, 2011; Setiono, Thong, & Yap, 1998; Taha & Ghosh, 1999).

The research detailed in the remaining chapters of this book concerns training artificial neural networks on musical tasks and then interpreting the internal structure of each trained network. We interpret networks in order to discover the manner in which they solve musical problems. Earlier we saw that Krumhansl (1990a) made a number of design decisions that guided her studies of musical cognition. Similar decisions guide the research described in the chapters that follow. Let us consider these design decisions.

Krumhansl (1990a) focused her experimental research by studying subjects' responses to musical pitch, typically building her stimuli from the manageable set of 12 pitch-classes. She did this because the principle of octave equivalence captures the notion that pitch-class is a psychologically valid concept, because pitch-class is the foundation of Western tonal music, and because combinations of pitch-classes can be used to define more complex musical entities such as intervals, chords, and scales.

The simulation research reported in this book also focuses on tasks that involve pitch-class representations of stimuli. One reason for this design decision is an endorsement of all of Krumhansl's (1990a) reasons for focusing on pitch. Later we demonstrate that pitch-class representations permit the definition of a number of interesting musical tasks. For instance, a network can be presented inputs represented in terms of constituent pitch-classes and can learn to perform such tasks as identifying a scale's tonic, determining whether a scale is major or minor, classifying various types of triads and tetrachords, and generating the next chord in a progression.

A second reason for emphasizing pitch-class representations in the current research is that a great deal of the theory of Western music is related to pitch-class (Forte, 1973). We will see many examples in which network interpretations relate to music theory, and this relationship is facilitated when training sets are represented using pitch-classes.

Krumhansl (1990a) also made a number of design decisions concerning her experimental methods, such as whether subjects required musical expertise, and what types of judgments were required of subjects. We make a number of analogous design decisions concerning the nature of the networks to study.

First, all of the simulation studies that I report involve supervised learning. That is, I train networks to generate a desired set of input/output responses. This is because my primary goal is to interpret the internal structure of trained networks. To accomplish this research goal it is extremely helpful to know precisely the responses that a network has learned.

Second, all of the tasks my networks learn via supervised training involve well-established concepts in Western music theory. One reason for this is that music theory itself is typically used to construct a set of training stimuli. A second reason is that for such tasks music theory itself is a powerful aid to network interpretation.

Third, all of the simulation studies that I report seek the simplest network architecture capable of learning a desired input/output mapping. For networks that include hidden units, this means finding the smallest number of required hidden units. If a network that had no hidden units can solve a problem, then I study this network. The reason for seeking the simplest networks that could learn a task is my goal of network interpretation: simpler networks are easier to interpret.

Fourth, many of the simulation studies that I report use a particular architecture that I call networks of value units (Dawson & Schopflocher, 1992). Such networks employ an activation function tuned so that processing units only turn on to a narrow range of incoming signals. One advantage of this architecture is that networks of value units have many desirable properties when the goal is to interpret their internal structure (Berkeley et al., 1995; Dawson, 1998, 2004, 2013). As the value unit architecture is not standard, the next section describes it in more detail, and explains its advantages for a research project that has as its goal the interpretation of the internal structure of trained networks.

2.4 The Value Unit Architecture

2.4.1 Activation Functions

A key element of a processor in an artificial neural network is its activation function. An activation function is a mathematical equation that converts a processor's net input into a numerical value called the processor's activity. If the processor is an output unit, then its activity is a response. If the processor is a hidden unit, then its activity is passed on as a signal to other processors in the network.

In modern artificial neural networks, most activation functions are nonlinear. The most common in the literature is the logistic function:

$$f(net) = \frac{1}{1 + e^{(-net+\theta)}}$$

In this equation, net is a processor's net input, and θ is the unit's bias. Figure 2-2 illustrates the logistic function. That it is nonlinear is evident in its sigmoid shape. The values of this function range from zero (when net input is at negative infinity) to one (when net input is at positive infinity). When net input is equal to θ, it produces an activity of 0.5. Thus, the bias is analogous to a processor's threshold. Processors that use the logistic activation function are called integration devices (Ballard, 1986).

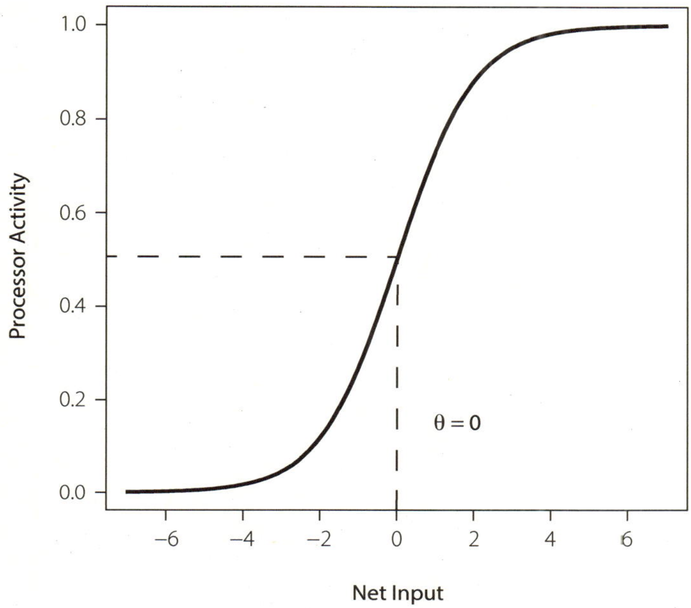

Figure 2-2 The logistic activation function used by an integration device to convert net input into activity.

The logistic activation function is the most common in connectionism and was fundamental to the discovery of learning rules for networks that include hidden units (Rumelhart et al., 1986). However, it is not the only activation function to be found in artificial neural networks. One review paper notes that an extremely large number of different activation functions exist in the connectionist literature (Duch & Jankowski, 1999).

One alternative activation function (Dawson & Schopflocher, 1992) uses a particular form of the Gaussian equation:

$$f(net) = e^{-\pi(net-\mu)^2}$$

In this equation, the value μ is analogous to the bias of an integration device. However, when net input equals μ this equation produces a maximum activity of one. As net input moves away from μ in either direction, activity drops quickly toward zero. Figure 2-3 illustrates the shape of this activation function. Because this function generates high activity to a very narrow range of net input values, processors that use this activation function are called value units (Ballard, 1986).

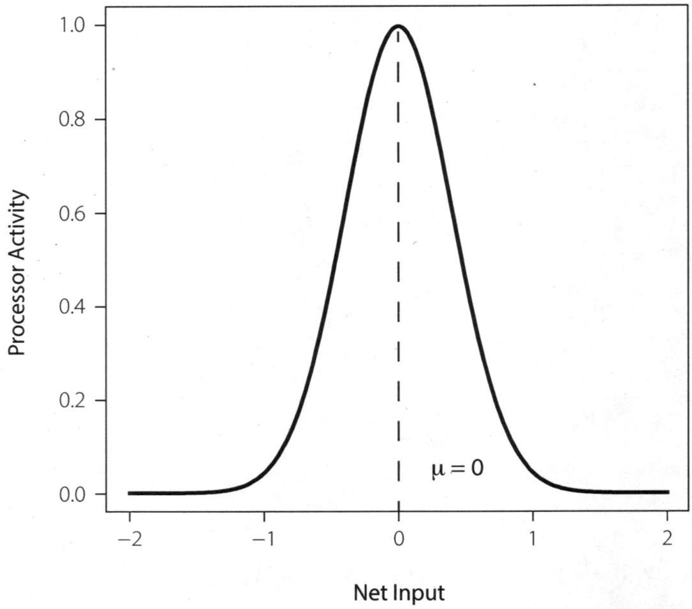

Figure 2-3 The Gaussian activation function used by a value unit to convert net input into activity.

One characteristic of an integration device is that after net input reaches a sufficiently high value, the activity that it is converted into is essentially "on." In Figure 2-2, there would not be an appreciable difference between activity when net input was 6 and activity when net input was 600. A value unit exhibits a very different sensitivity to net input. A value unit generates very high activity to a narrow range of net inputs and generates very low activity to any net input that is outside this narrow range. This different nature of a value unit's activation function often leads to advantages in comparison to the more traditional integration device architecture (Dawson & Schopflocher, 1992).

First, for many problems, networks of value units learn much faster than do networks of integration devices. Second, networks of value units tend to require fewer hidden units than do networks of integration devices when confronted with a complex problem.

Third, and most important in the context of this book, value units have emergent properties that make the internal structure of networks that contain them easier to interpret than networks of integration devices (Berkeley et al., 1995). In general, this is because each value unit is tuned, via its activation function, to respond to very particular combinations of stimulus features. This is not the case for integration devices, which in essence turn on when a sufficient number of stimulus features are present. The tuning that is inherent in the value unit architecture provides a window into how networks solve problems that is rarely available with the more traditional architecture. I have been able to take advantage of value unit properties to interpret the internal structure of many networks trained to solve problems in a wide variety of domains (Dawson & Boechler, 2007; Dawson, Boechler, & Orsten, 2005; Dawson, Boechler, & Valsangkar-Smyth, 2000a; Dawson, Medler, & Berkeley, 1997; Dawson, Medler, McCaughan, Willson, & Carbonaro, 2000b; Dawson & Piercey, 2001; Dawson & Zimmerman, 2003; Leighton & Dawson, 2001; Medler, Dawson, & Kingstone, 2005; Yaremchuk & Dawson, 2005, 2008).

This is not to say that network interpretation requires the value unit architecture. In some instances, networks of integration devices are indeed better for this task (Graham & Dawson, 2005). However, our experience has shown that value units are often advantageous when network interpretation is the goal, and it is for this reason that they are used in many of the simulations that we report. On the one hand, the success of these simulations provides more evidence in support of this network choice. On the other hand, the point of this book is the value of network interpretations; seeking interpretations in networks that use other activation functions is an important task that should be encouraged. Perhaps results like those reported in the chapters that follow will stimulate researchers to interpret the internal structure of other types of networks.

2.5 Summary and Implications

The musical cognitivism that was introduced in Chapter 1 is situated in classical cognitive science, which assumes that cognition results from the rule-governed manipulation of mental representations. Chapter 2 began by pointing out that alternative notions of cognition exist within cognitive science (Dawson, 2004). One of these is connectionism, which views cognition as emerging from non-symbolic information processing in the brain. Connectionist cognitive science models this type of information processing with artificial neural networks. Chapter 2 introduced some of the basic properties of these networks, including the properties of processing units, of weighted connections, and of the ability of these networks to learn from experience. It then provided an overview of the general methodological considerations that guided the simulations reported in the chapters that follow.

These include focusing on musical tasks that involve pitch-class representations of stimuli, the use of supervised learning, seeking the simplest networks capable of solving musical problems, and a preference for the value unit architecture. In general, the goal of the simulation research is to interpret the internal structure of networks trained on musical tasks in order to determine the kind of musical regularities that these networks exploit and represent.

2.5.1 Methodological Implications

The types of networks explored in the chapters that follow were initially developed in a tradition that explored spatial pattern recognition (McClelland & Rumelhart, 1986; Minsky & Papert, 1969; Pao, 1989; Ripley, 1996; Rosenblatt, 1962; Rumelhart & McClelland, 1986b; Schwab & Nusbaum, 1986). As a result, the networks typically learn to make judgments about sets of pitches that are presented simultaneously to a network—that is, across a spatial array of input units—instead of being presented in succession. These input representations are more closely related to the more abstract representation of music considered when mathematical set theory is applied to music (Forte, 1973, 1985; Roig-Francolí, 2008; Straus, 2005).

Choosing this type of input representation means that the musical tasks we consider involve classification (e.g., identifying a scale's type, classifying types of musical chords, identifying a composition's musical key, and so on). I will for the most part not be concerned with temporal properties of music, such as rhythm. I will also not explore temporal properties that involve presenting musical stimuli over time, note by note. However, it is important to realize that I am not in principle limited to analyzing non-temporal properties of music. For instance, later in this book we will see how a (spatial) network is presented an input chord, and then generates the next chord to be played in a particular progression. In addition, I could, in principle, present music temporally to these spatial networks. For instance, I could use a network's input units as a temporal window—an encoding of the pitches being "heard" at a particular moment in time—and then pass a musical stimulus over time through this input window.

By limiting the tasks below to those that I can present easily to our architectures of choice, I am simply exploring the possibility that such networks can provide new insights that may be relevant to musical cognition or to music theory. To the extent that I encounter success, I am motivated to explore in the future more complicated encodings to pursue similar temporal discoveries with the architectures described below. Furthermore, the exploration of music is not limited to the architectures that are employed in this book. Other architectures have been used (Griffith &

Todd, 1999; Todd & Loy, 1991), such as self-organizing networks that learn statistical properties of inputs without requiring an external teacher (Gjerdingen, 1990; Kohonen, 2001; Page, 1994) or recurrent networks that are explicitly designed to detect regularities in time (Elman, 1990; Franklin, 2004, 2006).

A key theme of this book is that regardless of the architecture that one uses to explore music with neural networks, or of the types of musical regularities being investigated, after training a network it is critically important to interpret its internal structure. The literature has long established the computational power of artificial neural networks, so the fact that a network can learn a particular task should not by itself be either interesting or surprising. The interesting information provided by networks can only emerge from their interpretation: finding out how networks actually solve the tasks that they learn (Dawson, 1998, 2004, 2009, 2013).

2.5.2 Synthetic Psychology

The methodological implications discussed in the previous section pertain directly to a connectionist study of music. Importantly, the networks to be presented in the chapters that follow reflect a more general research program called synthetic psychology (Braitenberg, 1984). In synthetic psychology, models are built first, and then used to produce data. The hope is that this data will include surprising phenomena, and that we will be in a position to offer straightforward accounts of these surprises because of our knowledge about the model itself.

It has been argued that one can use artificial neural networks in cognitive science to conduct synthetic psychology (Dawson, 2004). However, for this particular flavour of synthetic psychology to succeed, network interpretations must be supplied. It is never a surprise to find that some network can learn a particular task. This is because networks are, in principle, extremely powerful information processors (Siegelmann & Sontag, 1991). Following from this, one can only be surprised by the methods for solving problems that networks discover as they learn. Of course, to experience such surprises one must examine the internal structure of trained networks. From this perspective, one can consider this book a case study of how connectionists can perform synthetic psychology.

One consequence of using networks to advance synthetic psychology is that one aspect of connectionist Romanticism is not abandoned: the emphasis on individual networks. In some cases, such as when we examine how learning speeds are affected by different encodings, we train multiple networks on the same problem, treat each network as a different subject, and use statistics to get a sense of general performance. However, this is not the typical approach in this book. Instead, I will usually focus on a single network that has been trained on a particular problem, and interpret the internal structure of this particular subject.

There is, of course, a concern that this approach will not reveal average or typical network structures, because it focuses upon individuals instead of groups. However, my experience—particularly with the musical problems that I detail in the chapters that follow—is that this is not the case. I have examined the connection weights of many different networks trained on the same task as part of the research that has culminated in this book, and I repeatedly find similar internal structures in different networks. Indeed, I have taken advantage of this to use some of the tasks described later as exercises in a neural network course. Students in this course train networks in class and then find in their networks similar structures to those that follow. I am confident that my interpretations reflect typical network solutions to these musical problems. Of course, this does not rule out the interesting possibility that networks that use different internal structures to solve this problem can still be discovered.

2.5.3 Seeking New Music Theory

If the zeitgeist of the connectionist cognitive science of music is to capture that which cannot be formalized, then the general paradigm outlined above might seem odd. If we use pitch-class representations, if we use supervised learning, and if we train networks on established concepts of music theory, then what new information can we hope to learn? Should it not be the case that all we will pull out of our networks is the music theory that we put in?

Interestingly, this is not the case. My networks typically reveal alternative solutions to musical problems that lead to new ideas in music theory. The task of the remainder of this book is to provide evidence to support this claim. Let us begin by considering networks that are trained on a basic musical task: identifying the tonic note of a musical scale.

3

The Scale Tonic Perceptron

3.1 Pitch-Class Representations of Scales

In this chapter, I begin our synthetic psychology of music by training networks on a very basic task: identifying the tonic note of a stimulus scale. I begin by describing some basic properties of scales and consider how to present particular scales to a simple network and how to represent network responses to these stimuli. I then provide the details about the kind of network that I use and the method that I use to train it. I end by interpreting the internal structure of a trained network by examining the connection weights that are produced by training the network to generate the tonic of each scale in the training set.

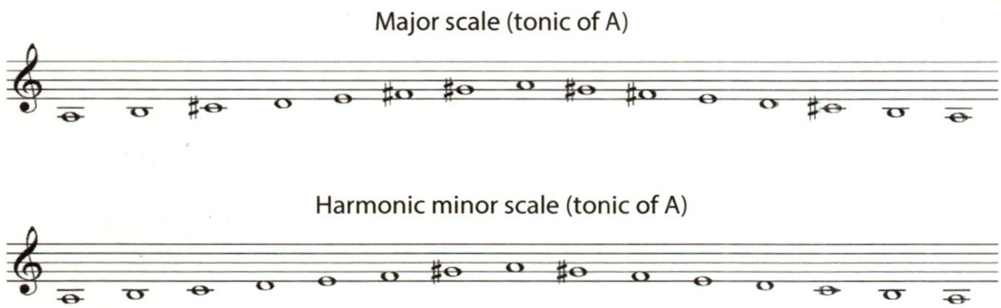

Figure 3-1 An example major scale, and an example harmonic minor scale, represented using multiple staffs.

3.1.1 Musical Scales

Musical scales provide the foundation of Western tonal music (Laitz, 2008). The current chapter is concerned with two of these scales: the major and the harmonic minor. Figure 3-1 provides an example of a major scale (ascending and descending) and an example of a harmonic minor scale (ascending and descending) in musical staff notation. Both of these scales are in the key of A. These two types of scales play a central role in defining the musical stimuli that I will train a simple artificial neural network to classify in this chapter.

When discussing a particular musical scale it is typical to identify each note in terms of its position or degree within the scale. Typically the tonic of the scale (its first note) is assigned the value 1, its second note is assigned the value 2, and so on (Laitz, 2008). Figure 3-1 labels each note in the two scales in this fashion. Each scale degree also has a name that is sometimes used instead of the note's number (Laitz, 2008). Note 1 is the tonic (or the root) of the scale, 2 is the supertonic, 3 is the mediant, 4 is the subdominant, 5 is the dominant, 6 is the submediant, and 7 is the leading-tone.

3.1.2 Scales and Pitch-Classes

What is the difference between a major scale and a harmonic minor scale? In Western tonal music, these scales differ in their patterns of distances or musical intervals between adjacent notes (Laitz, 2008). These different patterns in turn produce different musical experiences, sonorities, or emotional effects. In order to consider the patterns that define each of these scale types, let us first introduce the notion of pitch-class.

An examination of the two different musical scores in Figure 3-1 indicates that both use 18 notes to represent the scales. However, six of the notes (degrees 2 through 7) are repeated. One of the notes (degree 1, which is the note A in both scales) has two versions that are an octave apart.

In terms of musical sound, the two versions of the note A in the score are distinct. The first (the A below middle C on a piano, sometimes called A3) begins each scale and is associated with a note with a fundamental frequency of 220 Hz. The second (the A above middle C or A4) ends the first line of each scale and is associated with a note with a fundamental frequency of 440 Hz.

A3 and A4 are clearly different notes or different pitches; the former is lower than the latter. Phenomenologically, however, these two different notes seem quite related (Révész, 1954). "The octave note [A4] therefore bears a double relation to the prime tone [A3]. In one respect, of all the notes within the span of the octave, it is the one most dissimilar to the prime tone, since it is the farthest from it in point of distance. In another respect, however, it is the note most similar to the prime tone, since of all the notes through which we passed, it is the one most similar to it in quality" (Révész, 1954, pp. 56–57). This relationship between notes an octave apart is called octave equivalence.

Révész (1954) used the phenomenological notion of octave equivalence to argue that music cognition cannot simply pay attention to physical properties (e.g., sound frequency) but must also be sensitive to a nonphysical dimension: musical structure. Psychophysical experiments inspired by this perspective have studied octave equivalence in a variety of species, including humans, rats, starlings, and chickadees (Allen, 1967; Blackwell & Schlosberg, 1943; Cynx, 1993; Demany & Armand, 1984; Hoeschele, Weisman, Guillette, Hahn, & Sturdy, 2013; Shepard, 1964). These studies provide mixed support for octave equivalence. Some results suggest that the demonstration of octave equivalence depends critically on musical context and listener expectations (Deutsch & Boulanger, 1984). Others claim that octave equivalence may be a universal property (Patel, 2008).

While its psychophysical support is murky, octave equivalence is a central assumption in the formal analysis of music (Forte, 1973; Hanson, 1960; Lewin, 2007; Roig-Francolí, 2008; Straus, 1991, 2005; Tymoczko, 2011). Musical analysis assigns two pitches separated by one or more octaves (e.g., A3 and A4) to the same pitch-class (i.e., A). In so doing, it uses pitch-class set notation, sometimes shortened to pc set, to provide a simpler representation of the notes presented in standard staff notation like Figure 3-1. With pc sets, a musical selection becomes a set of pitch-classes. For instance, while both scales in Figure 3-1 are written in the staves with 18 notes, in reality each only employs seven different pitch-classes, which is why scale degree can be represented using only seven different Roman numerals.

Forte (1973) uses the axiom of octave equivalence to assign two different pitches that are an octave apart to the same pitch-class. In Western music, there are only 12 different pitch-classes available (see Figure 3-2 below). Forte further simplifies pitch-class set notation with the axiom of enharmonic equivalence. In standard staff notation, different symbols can represent the same pitch. For instance, A4 is a particular musical pitch. In a musical score, depending upon context, this one pitch could be represented with the symbol "A," the symbol "B♭♭," or the symbol "G♯♯." The axiom of enharmonic equivalence uses one symbol (A) to stand for any of these different symbols. (In this book, this axiom is exploited to reduce the use of the ♭ symbol.)

Let us illustrate the structure of the major and harmonic scales from Figure 3-1 with a geometric exploitation of pitch-class notation. First, we arrange all 12 pitch-classes of Western music around a circle of minor seconds, so that adjacent pitch-classes are a minor second (one semitone) apart. Second, we draw spokes from the centre of the circle to indicate which pitch-classes are included in a scale. Figure 3-2 provides this illustration for both of the scales from Figure 3-1. Note that both figures have only seven spokes, because both scales are constructed using only seven different pitch-classes.

Western music is typically tonal: compositions are set in a particular musical key, which in turn provides important structure to an audience. For instance, when a musical key is established, some notes (in particular the tonic, subdominant, and dominant notes of the key) are more stable than others (Krumhansl, 1990a). As a result, a composer can manipulate a piece's effect on a listener by moving from less stable to more stable tonal elements (Schoenberg, 1969). Tonality is created in music by restricting the use of musical notes (Browne, 1981). Consider the pitch-classes of the A major scale illustrated in Figure 3-2. In the circle of minor seconds, there are 12 different pitch-classes available. However, the A major scale only employs seven of them: A, B, C♯, D, E, F♯, and G♯.

The pitch-classes that are not used in the A major scale are missing for a reason. A major scale has a specific pattern of distances between adjacent notes. Again, consider Figure 3-2's illustration of A major. Start at A, and move clockwise around the circle. The next note with a spoke (B) is two semitones, or a full tone, higher than A. The next note encountered (C♯) is a full tone higher than B. However, the next note with a spoke (D) is only a semitone higher than the preceding note (C♯). Following the circle all the way around until the pitch-class A is encountered again, a specific pattern of between-note distances is apparent: tone-tone-semitone-tone-tone-tone-semitone. This pattern of between-note distances defines a major scale (Laitz, 2008).

If one excludes a different subset of pitch-classes from the circle of minor seconds, then one creates a different type of musical scale. Consider the A harmonic minor scale that is also represented in Figure 3-2. The A harmonic minor scale is defined by the pitch-classes A, B, C, D, E, F, and G♯, which produces different between-note distances. Look at the A harmonic diagram in Figure 3-2, starting at A and moving clockwise around the circle. Most of the distances are either semitones or full tones as was the case for A major, but there is also one distance (from F to G♯) that is an augmented second (three semitones). The complete pattern of between-note distances that defines a harmonic minor scale is tone-semitone-tone-tone-semitone-augmented second-semitone.

Importantly, the pattern of between-note distances that defines either a major or a harmonic minor scale is constant. This means that the two spoke patterns presented in Figure 3-2 define the major or harmonic scale whose tonic is any of the 12 pitch-classes. For instance, consider the spokes of the A major circle as a solid unit. If we rotate this unit 30° clockwise, then the between-note distances will be unchanged, but the pattern starts on a new tonic note (B♭). This rotated spoke pattern defines the B♭ major scale. Similarly, if one takes the entire set of spokes for the A harmonic minor scale in Figure 3-2 and rotates them 30° counter-clockwise, the pattern now defines the G♯ harmonic minor scale.

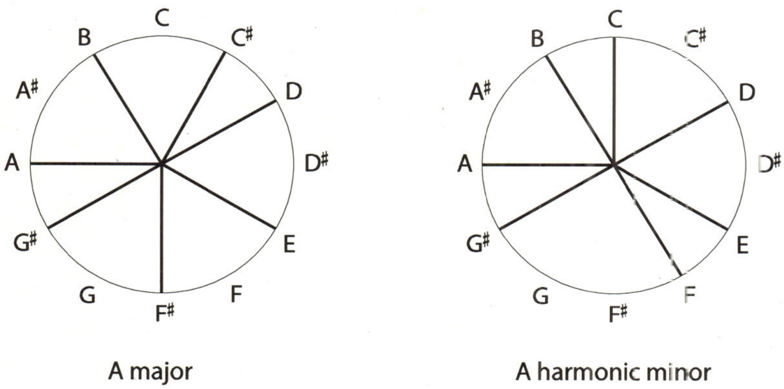

Figure 3-2 The circle of minor seconds can be used to represent the pitch-classes found in the A major and the A harmonic minor scales.

The spoke patterns illustrated in Figure 3-2 clearly represent any major or any harmonic minor musical scale. This suggests, for instance, that if we were presented with a particular spoke pattern, then we could examine it for particular patterns of between-note distances, identifying whether a major or harmonic minor scale was depicted. Furthermore, a closer examination of the relative positions of the various spokes would permit us to determine the tonic note of the depicted scale.

This latter task defines our first case study of a musical network. I will present a perceptron the set of pitch-classes that define a major or harmonic minor scale, and train it to identify the tonic note of the presented scale.

3.1.3 Pitch-Class Representations

Pitch-class representations permit mathematics to be used to explore and manipulate musical structure (Forte, 1973; Lewin, 2007). For mathematical manipulation, a pitch-class representation is an ordered set of numbers that indicate which pitch-classes are present in a musical entity. Using Allen Forte's pitch-class notation, each pitch is assigned a specific integer, with C represented as 0, C♯ as 1, and so on. Therefore, in this system, the raw pitch-class representation of the A major scale is the ordered set [9, 11, 1, 2, 4, 6, 8].

Pitch-class representations are also often used when artificial neural networks learn to classify musical patterns. However, this sort of pitch-class representation is slightly different from the mathematical representation described above. For pitch-class representation in a musical network, the network has 12 different input units. Each input unit represents the presence or absence of a particular pitch-class, as is the case in Figure 3-3 in Section 3.2. For instance, imagine that the first input unit represents the pitch-class A; the second input unit represents the pitch-class B; and so on. If the pitch-class A is present in a stimulus, then we turn its input unit "on" by giving it an activity of one. However, if the pitch-class A is absent from a stimulus, we turn its input unit "off" using an activity of zero. In short, a pitch-class representation for a network is a string of 12 bits representing the presence or absence of the 12 possible pitch-classes. For instance, one would use the following pitch-class representation to present the A major scale to a network: [1, 0, 1, 0, 1, 1, 0, 1, 0, 1, 0, 1].

Table 3-1 provides this type of pitch-class representation for each of the possible major and harmonic minor scales in Western music. This table defines 24 different stimuli that can be presented to a perceptron that has 12 different input units. In the next section, I describe training a perceptron to accomplish the following pattern recognition task: when presented with one of the stimuli provided in Table 3-1, the perceptron will indicate the tonic note of that stimulus, ignoring whether the stimulus represents a major or a harmonic minor scale.

Before proceeding with an account of the perceptron, we can use Table 3-1 to illustrate why this task is not going to be straightforward for a perceptron to learn. We saw in Figure 3-2 that harmonic minor scales are distinct in that they include an interval of an augmented second. One might think that a perceptron could simply detect that interval and then use its position to determine the scale's tonic. However, the easy identification of the augmented second requires that the pitch-classes already be arranged in the order in which they are found in a harmonic minor scale. This will not be the case for the perceptron; it will always receive pitch-classes in the same order regardless of scale. That is, the input patterns are the different rows of Table 3-1. One cannot simply look at any of those rows and immediately see where the augmented second lies, or for that matter where the tonic of the scale is positioned. I am making the problem difficult for the perceptron by always presenting pitch-classes in the same order. For this reason, I am interested in discovering how the perceptron deals with this difficulty when it learns how to identify scale tonics.

Table 3-1 Pitch-class representation (for an artificial neural network) of 12 different major and 12 different harmonic minor scales.

	Scale tonic	Pitch-class components of scale											
		A	A#	B	C	C#	D	D#	E	F	F#	G	G#
Major scale	**A**	1	0	1	0	1	1	0	1	0	1	0	1
	A#	1	1	0	1	0	1	1	0	1	0	1	0
	B	0	1	1	0	1	0	1	1	0	1	0	1
	C	1	0	1	1	0	1	0	1	1	0	1	0
	C#	0	1	0	1	1	0	1	0	1	1	0	1
	D	1	0	1	0	1	1	0	1	0	1	1	0
	D#	0	1	0	1	0	1	1	0	1	0	1	1
	E	1	0	1	0	1	0	1	1	0	1	0	1
	F	1	1	0	1	0	1	0	1	1	0	1	0
	F#	0	1	1	0	1	0	1	0	1	1	0	1
	G	1	0	1	1	0	1	0	1	0	1	1	0
	G#	0	1	0	1	1	0	1	0	1	0	1	1
Harmonic minor scale	**A**	1	0	1	1	0	1	0	1	1	0	0	1
	A#	1	1	0	1	1	0	1	0	1	1	0	0
	B	0	1	1	0	1	1	0	1	0	1	1	0
	C	0	0	1	1	0	1	1	0	1	0	1	1
	C#	1	0	0	1	1	0	1	1	0	1	0	1
	D	1	1	0	0	1	1	0	1	1	0	1	0
	D#	0	1	1	0	0	1	1	0	1	1	0	1
	E	1	0	1	1	0	0	1	1	0	1	1	0
	F	0	1	0	1	1	0	0	1	1	0	1	1
	F#	1	0	1	0	1	1	0	0	1	1	0	1
	G	1	1	0	1	0	1	1	0	0	1	1	0
	G#	0	1	1	0	1	0	1	1	0	0	1	1

Note. Each row provides the pitch-classes that are included in a particular scale whose mode and tonic are provided in the two columns on the left. The number 1 indicates that a pitch-class is included in the scale, and a 0 indicates that it is not included in the scale.

3.2 Identifying the Tonics of Musical Scales

3.2.1 Task

Our goal is to train a perceptron to determine the tonic note of a pattern of pitch-classes presented to its input units; each presented pattern defines either a major scale or a harmonic minor scale.

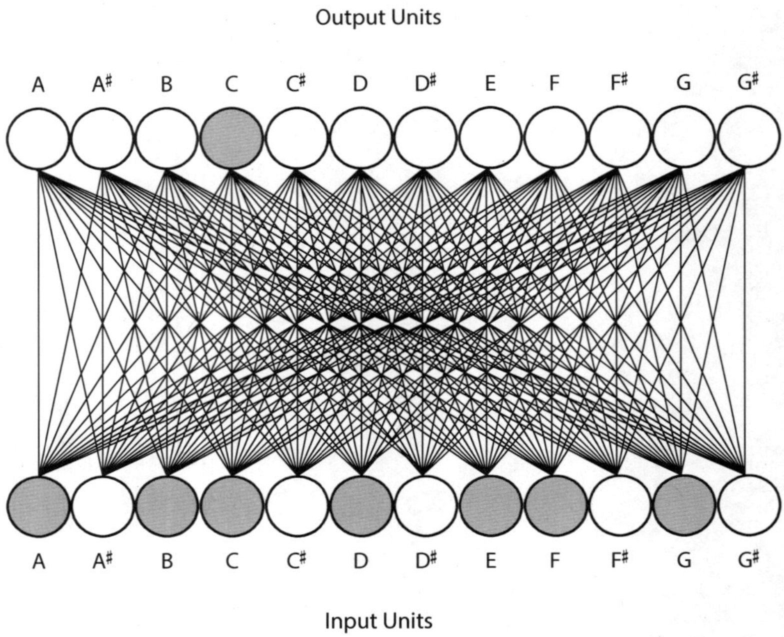

Figure 3-3 Architecture of a perceptron trained to identify the tonic notes of input patterns of pitch-classes.

3.2.2 The Perceptron

A perceptron is a simple artificial neural network originally invented by Frank Rosenblatt in the 1950s (Rosenblatt, 1958, 1962). This type of network consists of a layer of input units directly connected to a layer of output units (Figure 3-3). In this figure, circles represent processing units and lines between circles represent weighted connections between processers through which signals are sent. Perceptrons are simple because they do not include any hidden units between input and output units. This means that perceptrons are less powerful than the more modern multilayer networks that we will see in the next chapter (Minsky & Papert, 1969).

However, perceptrons are still powerful enough to provide interesting models of some aspects of cognition (Dawson, 2008; Dawson, Dupuis, Spetch, & Kelly, 2009; Dawson, Kelly, Spetch, & Dupuis, 2010b).

How do perceptrons learn? Modern artificial neural networks use continuous, nonlinear activation functions to convert net inputs into output unit responses. As indicated in Chapter 2, many of the networks described in this book use value units, which employ a Gaussian activation function. The use of a continuous activation function permits networks to be trained using gradient descent learning. Gradient descent learning is a form of supervised learning in which output unit error is used to modify existing connection weights in order to reduce future error. That is, after connection weights are changed, the next time the same pattern is presented to the network it will generate less error from it. In general, learning is defined as $w_{ij}(t+1) = w_{ij}(t) + \Delta w_{ij}$, where w_{ij} is the weight of the connection from input unit i to output unit j, (t+1) and (t) indicate time, and Δw_{ij} is a computed weight change. This equation simply says that each new weight is equal to some computed weight change that is added to the old weight.

The computed weight change is based on the error calculated for the output unit. In general, it is equal to a fractional learning rate (ε) times the error computed for the unit at the output unit end of the connection $(t_j - a_j)$ times the activity of the unit at the input end of the connection (a_i). In other words $\Delta w_{ij} = \varepsilon(t_j - a_j)a_i$. In gradient descent learning, output unit error is also scaled by the derivative of the (continuous) activation function (Dawson, 2004, 2008). This attempts to help learning by changing weights in such a way that the network moves down the steepest slope of an error surface. The particular mathematics of gradient descent learning depends on which activation function is used in the network (Dawson & Schopflocher, 1992).

By repeatedly presenting each of the patterns in a training set, and by modifying connection weights using supervised learning, I can reduce errors to the point that it can be said that the perceptron is generating the correct response to each stimulus. At this point, one can say that the network has converged. That is, it has discovered a set of connection weights that correctly converts each stimulus into a correct response.

3.2.3 The Scale Tonic Perceptron

The scale tonic perceptron, illustrated in Figure 3-3, has 12 input units, each representing the presence or absence of a particular pitch-class in a stimulus. The perceptron also has 12 output units, which also represent different pitch-classes. The input units are used to present the perceptron a scale represented as pitch-classes (any of the rows in Table 3-1). The perceptron learns to turn on only one of its output units, the one that represents the pitch-class of the tonic of the input pattern. The grey shading in Figure 3-3 illustrates an example of a desired outcome of training: the network correctly responds with a tonic of C when presented the pitch-classes of the C major scale. Each output unit in the perceptron is a value unit that uses a Gaussian activation function to convert incoming signals to internal activity (Dawson & Schopflocher, 1992).

3.2.4 Training Set and Training

The training set consists of 24 different input patterns; each pattern is a row of numbers taken from Table 3-1. For any input pattern, the desired response turns the output unit representing the pattern's tonic note on, and turns the other 11 output units off. Training is conducted with the gradient descent rule developed specifically for perceptrons whose output processors are value units (Dawson, 2004). The software for training the perceptron was the Rosenblatt program (Dawson, 2005), which is available as freeware from the author's website http://www.bcp.psych.ualberta.ca/~mike/AlienMusic/

All connection weights in the perceptron are set to random values between −0.1and 0.1 before training begins. The biases of the output units (i.e., the value of μ for their Gaussian activation functions) are initialized to zero, and are not modified by training. We employ a learning rate of 0.005. Training proceeds until the network generates a "hit" for each output unit, for each of the 24 patterns in the training set. A "hit" is defined as activity of 0.9 or higher when the desired response is one, or as activity of 0.1 or lower when the desired response is zero.

Under these conditions, perceptrons of the type represented in Figure 3-3 learn to identify the tonic of an input pattern extremely rapidly, generally requiring between 20 and 30 epochs of training before generating 12 "hits" to each training stimulus. An epoch of training for this network involves training the network once on each of the 24 training patterns; we randomize the order of pattern presentation every epoch. The particular network described in more detail in the next section learned to solve the problem after only 19 epochs of training.

3.3 Interpreting the Scale Tonic Perceptron

3.3.1 Interpreting Connection Weights

A perceptron consists simply of a set of input units connected to a set of output units. Therefore, to interpret the internal structure of this type of network one is limited to examining the various connection weights in the network after training has succeeded. Table 3-2 above provides the entire set of 144 connection weights observed in a typical scale tonic perceptron at the end of training. Each column of the table is associated with one of the output units; each row of the table is associated with one of the input units. Therefore, one column of numbers in Table 3-2 provides the set of connection weights from each of the 12 input units to one of the output units.

Table 3-2 The connection weights from each input unit to each output unit for a perceptron trained to identify the tonic pitch-class of an input major or harmonic minor scale pattern.

Input units (scale pitch-classes)	Output units (tonic pitch-classes)											
	A	A#	B	C	C#	D	D#	E	F	F#	G	G#
A	0.06	−0.24	0.75	0.38	−0.40	0.33	−0.65	0.27	−0.30	−0.18	−0.11	0.44
A#	0.37	0.10	−0.34	0.77	−0.49	−0.49	0.29	−0.60	0.19	−0.16	−0.24	0.02
B	−0.01	0.36	0.07	−0.25	−0.72	−0.44	−0.55	0.32	−0.65	0.18	−0.34	0.13
C	0.27	0.01	0.41	−0.05	0.26	−0.63	−0.41	−0.52	0.37	−0.77	0.18	0.22
C#	0.39	0.22	0.04	0.48	−0.10	0.37	−0.77	−0.47	−0.49	0.38	−0.71	−0.09
D	−0.20	0.29	0.21	0.19	−0.36	−0.15	0.33	−0.75	−0.47	−0.48	0.44	0.80
D#	0.68	−0.24	0.33	0.20	0.00	−0.44	−0.04	0.26	−0.75	−0.5[illegible]	−0.40	−0.38
E	−0.30	0.63	−0.21	0.32	−0.33	−0.07	−0.29	−0.08	0.25	−0.75	−0.34	0.49
F	0.54	−0.36	0.65	−0.17	−0.26	−0.23	−0.04	−0.33	−0.09	0.25	−0.75	0.45
F#	0.45	0.49	−0.33	0.80	0.12	−0.32	−0.22	−0.03	−0.39	−0.05	0.29	0.75
G	0.74	0.46	0.52	−0.37	−0.74	0.23	−0.41	−0.23	−0.05	−0.46	−0.15	−0.33
G#	−0.31	0.81	0.45	0.47	0.44	−0.64	0.23	−0.30	−0.22	−0.12	−0.33	0.16

Note. Each row provides the connection weight from a particular input unit to each of the 12 output units.

The set of connection weights in Table 3-2 represents one perceptron's "knowledge" about the relationship between scale patterns and tonic pitch-classes. The problem is to discover the nature of this knowledge by examining this entire pattern of connectivity. Ordinarily, when confronted with a matrix of numbers like Table

3-2, one might explore its structure by using multivariate statistics (such as factor analysis or multidimensional scaling) to make the matrix more understandable. Fortunately, we can understand the workings of a scale tonic perceptron by simply rearranging Table 3-2 in a manner that takes advantage of some general, well-known musical properties.

Consider the first column of Table 3-2. The first entry in this column is the weight from the input pitch-class A to the output tonic pitch-class A. However, this entry can also be described as the weight of the connection to the output pitch-class A from the pitch-class that is zero semitones away from the output pitch-class (moving in a clockwise direction around the circle of minor seconds that was used in Figure 3-2). Similarly, the second entry in the first column of Table 3-2 is also the weight to that pitch-class from the note that is one semitone away around the circle of minor seconds.

When we apply this alternative interpretation to other columns of Table 3-2, we find that entries that are in the same row do not represent notes the same distance away from the output pitch-class. For instance, the first entry in the second column is the weight to the output pitch-class A♯ from the input pitch-class (A) that is 11, not zero, semitones away.

One can use this alternative interpretation to rearrange the Table 3-2 weights so that the first connection weight appearing in a column corresponds to a zero semitone distance between the input and output pitch-classes, the next corresponds to a one semitone distance, and so on. Table 3-3 represents the Table 3-2 weights in this alternative manner. Importantly, the connection weights in Table 3-3 are identical to those in Table 3-2; we simply provide them a different row label and reorganize the table to reflect the difference in labelling. In a real sense, the perceptron itself "interprets" the weights as labelled in Table 3-3, and not as labelled in Table 3-2, because an inspection of Table 3-3 reveals an elegant solution to converting an input scale pattern to an output tonic pitch-class, as we will now proceed to discuss.

If one inspects the values of the connection weights along each row of Table 3-3 (ignoring for the moment whether the weight is positive or negative), then one sees commonality. In general, all of the connection weights in each row are roughly the same size. (If one ignores sign, and computes the standard deviation for each row, the standard deviations range in size from 0.04 to 0.07.) In other words, if one considers input pitch-classes in terms of relative distance from the output pitch-class rather than in terms of absolute pitch-class, then structure emerges without the need for multivariate statistics.

Table 3-3 The rearranged connection weights from Table 3-2.

Number of semitones between input pitch-class and output pitch-class	Output units (tonic pitch-classes)											
	A	A#	B	C	C#	D	D#	E	F	F#	G	G#
0	0.06	0.10	0.07	−0.05	−0.10	−0.15	−0.04	−0.08	−0.09	−0.05	−0.15	0.16
1	0.37	0.36	0.41	0.48	−0.36	−0.44	−0.29	−0.33	−0.39	−0.46	−0.33	0.44
2	−0.01	0.01	0.04	0.19	0.00	−0.07	−0.04	−0.03	−0.05	−0.12	−0.11	0.02
3	0.27	0.22	0.21	0.20	−0.33	−0.23	−0.22	−0.23	−0.22	−0.18	−0.24	0.13
4	0.39	0.29	0.33	0.32	−0.26	−0.32	−0.41	−0.30	−0.30	−0.16	−0.34	0.22
5	−0.20	−0.24	−0.21	−0.17	0.12	0.23	0.23	0.27	0.19	0.18	0.18	−0.09
6	0.68	0.63	0.65	0.80	−0.74	−0.64	−0.65	−0.60	−0.65	−0.77	−0.71	0.80
7	−0.30	−0.36	−0.33	−0.37	0.44	0.33	0.29	0.32	0.37	0.38	0.44	−0.38
8	0.54	0.49	0.52	0.47	−0.40	−0.49	−0.55	−0.52	−0.49	−0.48	−0.40	0.49
9	0.45	0.46	0.45	0.38	−0.49	−0.44	−0.41	−0.47	−0.47	−0.51	−0.34	0.45
10	0.74	0.81	0.75	0.77	−0.72	−0.63	−0.77	−0.75	−0.75	−0.75	−0.75	0.75
11	−0.31	−0.24	−0.34	−0.25	0.26	0.37	0.33	0.26	0.25	0.25	0.29	−0.33

Note. In this table each row indicates the relative distance between an input unit's pitch-class and an output unit's pitch-class measured in semitones. Thus the connection weights across a row are not from the same input unit, but are instead from different input units that serve the same role in different scales.

Alternatively, the common pattern in each row suggests that any column of weights from Table 3-3 represents a pattern of connectivity that is constant for any output unit. Indeed, this is true for output pitch-classes C♯, D, D♯, E, F, F♯, and G. The remaining output pitch-classes (A, A♯, B, C, and G♯) have essentially the same pattern, but the pattern has been multiplied by −1. Importantly, this difference (being inverted by multiplication by −1) has no effect on output unit behaviour, because the Gaussian activation is symmetric in the positive and the negative directions away from μ.

Recognizing the irrelevance of "column sign," we can take the connection weights in the columns for the inverted output pitch-classes (A, A♯, B, C, and G♯) and multiply them by –1 to make them coincide with the pattern found in the other seven columns. Then, we can compute the average of each of the 12 rows in Table 3-3 in order to determine the average pattern of connectivity from the input units to the output units. Figure 3-4 provides this average pattern of connectivity.

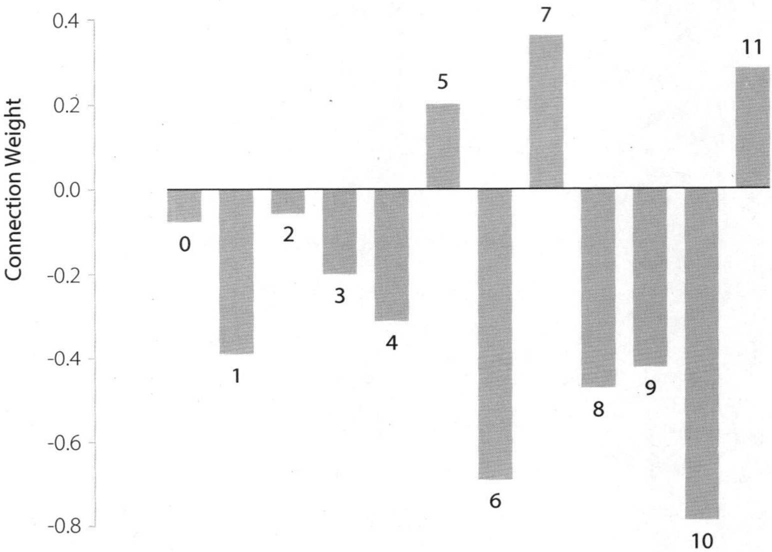

Figure 3-4 The connection weights between the 12 input units and any output unit in the scale tonic perceptron.

On first inspection, it is not immediately apparent how the pattern of weights in Figure 3-4 permits the perceptron to identify an input pattern's tonic. It is interesting, from a musical perspective, that the most extreme weights are from the pitch-classes that are either 6 or 10 semitones away from the output pitch-class. This is because both of these pitch-classes are absent from both the major and the minor harmonic scales. What is the relationship between the other weights and the tonic?

We answer these questions by undertaking a more specialized inspection of the connection weight pattern. Figure 3-5 plots the same weights as depicted in Figure 3-4 but removes the bars representing weights that receive a zero signal when a major scale pattern is input. Only seven bars remain, because only seven input units are turned on. One key observation about Figure 3-5 is the spacing between adjacent bars. Note that the spacing between bars, measured in tones and semitones, is exactly the spacing that we saw earlier described a major scale pattern.

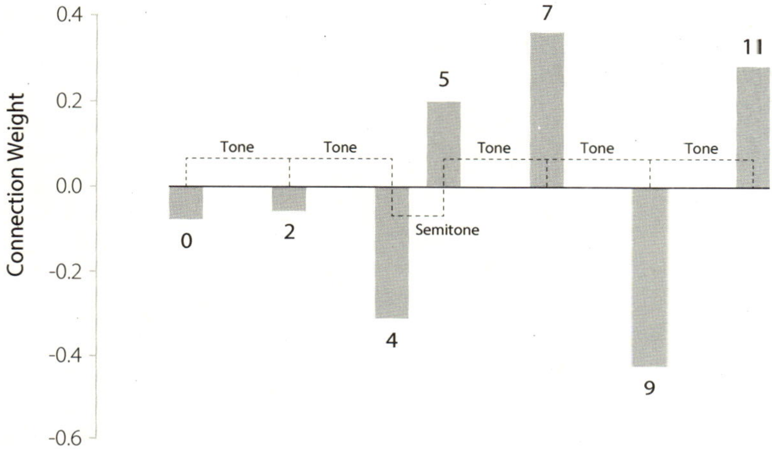

Figure 3-5 The connection weights between the 12 input units and an output unit in the scale tonic perceptron, showing only those connections that have a signal sent through them when the output unit's major scale pattern is presented to it.

A second property of the weights plotted in Figure 3-5, which is not quite as evident from the figure, is that if one sums their seven values, the result is −0.04. (The sum of these weights is identical to the net input sent to this output unit when the major scale associated with it is presented.) This is important because when the perceptron was trained, the value of μ in each output unit's Gaussian activation function was held constant at zero. This means that in order to generate a maximum output response, the net input flowing into the output unit must be near zero.

A value of −0.04 is close enough to zero to generate Gaussian activity that is near maximum (0.9950). In short, if signals are sent through the seven connection weights plotted in Figure 3-5—as happens when the major scale pattern associated with that output unit is presented to the network—then the output unit of the tonic pitch-class for this major scale will activate, producing a correct response.

A similar account holds for the case in which the network's stimulus is the harmonic minor scale. Figure 3-6 is similar to Figure 3-5, but in this case only plots the connection weights that carry signals from the harmonic minor scale input pattern. Again, the distances between adjacent bars are identical to the harmonic minor scale pattern that was discussed earlier, and the sum of the weights for the net input to the output unit) is close enough to zero (i.e., −0.01) to generate near maximum output unit activity (0.9997).

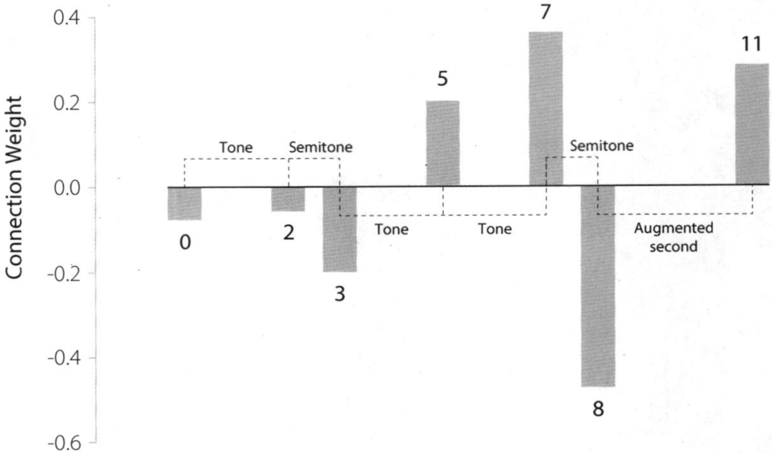

Figure 3-6 The connection weights between the 12 input units and an output unit, showing only those connections that have a signal sent through them when the output unit's harmonic minor scale pattern is presented to it.

In short, the pattern of weights in Figure 3-4 is highly structured. For the particular output unit that receives signals through these weights, the pattern of weights turns the output unit on when we present the major scale or harmonic minor scale built on the output unit's tonic. Importantly, for some other stimulus—for instance, the major scale that is associated with a different output unit's tonic pitch-class—this output unit will not turn on. Because the signal coming from the input units will not match this unit's major or harmonic minor scale, the net input will not be close to zero, and therefore the output unit will generate near-zero activity. For instance, imagine that Figures 3-4, 3-5, and 3-6 all represent the pattern of connectivity from the input pitch-classes to output pitch-class A. For this output unit, the bar labelled 0 comes from the input pitch-class A, the bar labelled 1 comes from the input pitch-class A♯, and so on.

Now imagine that we present the G major scale to the perceptron. This involves turning on a particular set of input units: A, B, C, D, E, F♯, and G. However, if we consider these input pitch-classes in terms of the relative distance not to G but instead to the output unit for A, we get a very different pattern of incoming signals from either Figure 3-5 or 3-6.

Figure 3-7 plots the connections to output unit A that are active for the G major scale stimulus. This pattern does not match either the major scale pattern of Figure 3-5 or the harmonic minor scale pattern of Figure 3-6. Indeed, if one sums the Figure 3-7 bars, the result (the net input to output unit A) is –1.00. This value is so far away from the value of μ (which, again, equals zero) that the output unit for A generates a very weak Gaussian activity (0.0432).

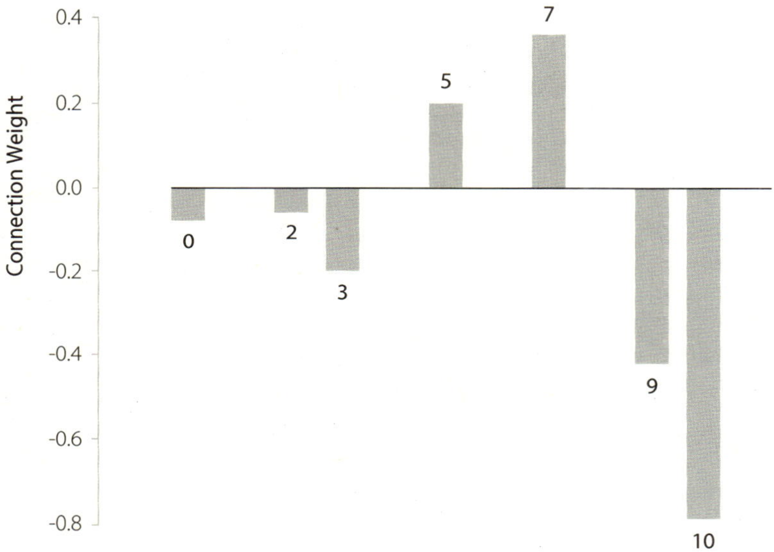

Figure 3-7 The active connections to the output unit for pitch-class A when the network is presented the G major scale.

3.3.2 Summary of the Perceptron

We can now summarize what we know about the artificial neural network that has been the subject of this chapter.

We trained a perceptron (Figure 3-3) to identify the tonic pitch-class of a major or harmonic minor scale presented to the network. The perceptron learned this task very quickly, in 19 epochs of training. An examination of the 144 connection weights in the perceptron revealed that it had quickly discovered some key musical properties that it exploited to accomplish this task.

First, the perceptron learned not to treat input units as if they represented absolute pitch-classes. Instead, the perceptron "considered" input units in terms of their semitone distance (clockwise around the circle of minor seconds) from the pitch-class represented by each output unit. For each output unit, the perceptron found the input pitch-class zero semitones away, and set the weight between the input and output unit to be roughly equal to the first bar in Figure 3-4. Similarly, it found the input pitch-class one semitone away, and assigned the connection between it and the output unit to be roughly equal to the second bar in Figure 3-4; this process was repeated for all of the remaining input units.

The resulting pattern of connection weights (Figure 3-4) for each output unit pitch-class combined the two scale patterns discussed in Section 3.1.2 into one set of connection weights. If the pattern of pitch-classes presented to the perceptron is the major scale for a particular output unit, then signals sent through the connection weights come from pitch-classes spaced at distances tone-tone-semitone-tone-tone-semitone away from the tonic (Figure 3-5). This produces a net input near zero, which activates the output unit.

Similarly, if the pattern of pitch-classes presented to the perceptron is the harmonic minor scale for a particular output unit, then signals sent through the connection weights come from pitch-classes spaced at distances tone-semitone-tone-tone-semitone-augmented second-semitone (Figure 3-6). Again, this produces a near-zero net input, which turns the correct output unit on.

Importantly, the pattern of connection weights in Figure 3-4 is such that major or harmonic minor scale inputs turn the correct output unit on, but also turn the remaining output units off. This is because these patterns produce net inputs to the other 11 output units that are sufficiently different from μ to generate near-zero activity.

In short, the perceptron quickly solves the scale tonic problem by discovering the between-distance spacing of pitch-classes that define both the major and the harmonic minor scales. What is astonishing is that this basic musical knowledge—a component of basic musical training (Martineau, 2008)—is acquired by a learning rule in which perceptron weights are slightly adjusted after each pattern presentation in order to decrease the measured error in each output unit. It may also be surprising to realize that the confusing array of connection weights presented in Table 3-2 represent this basic musical knowledge.

3.4 Summary and Implications

In this chapter, I described the training of a network that generated the appropriate scale tonic when presented with the pitch-classes that defined a particular major or harmonic minor scale. The resulting connection weights reflected basic music theory, as these weights captured the between-note distances that characterize both types of scales.

It is perhaps surprising that a single set of connection weights can identify the tonic of both a major and a harmonic minor scale. Even more interesting is the fact that the perceptron uses the same pattern of weights—rotated to correspond to the appropriate pitch-classes—to identify the tonics of different scales.

Given that a perceptron quickly learns to identify a scale pattern's tonic, one might hypothesize that a similar network could learn to classify an input scale pattern's mode—that is, to identify whether an input pattern represents a major or minor key. A perceptron for this task would use 12 input units to represent input pitch-classes, and a single output unit to represent mode. This perceptron could be presented any of the 24 stimuli that were presented in Table 3-1. We could try to train this network to turn this output unit on if the presented pattern was a major scale, and to turn this output unit off if the presented pattern was a harmonic minor scale.

At face value, identifying scale mode seems simpler than identifying scale tonic. However, and under a wide variety of training conditions (e.g., different learning rates, different initial conditions for weights), one discovers that a perceptron cannot be taught to accomplish this task. It could never generate a correct response to each of the 24 patterns in the training set.

This illustrates an important property of perceptrons: because of their simple structure (i.e., because they have no intermediate processors between input and output units), they are not powerful enough to solve every pattern recognition problem (Minsky & Papert, 1969). The reason for this is that without additional processors, called hidden units, they cannot detect higher-order regularities that are required to solve complex problems (Dawson, 1998, 2004, 2013).

The fact that a network as simple as a perceptron is capable of solving the scale tonic problem is highly informative, because it reveals that this particular musical problem is computationally simple. That a perceptron cannot detect whether a scale is major or minor indicates that a more powerful type of network, called a multilayer perceptron, is required for other musical problems. The next chapter introduces the multilayer perceptron and provides a case study wherein it is used to explore music by training it to detect a scale's mode: whether a presented pattern is associated with a major or harmonic minor scale.

4

The Scale Mode Network

4.1 The Multilayer Perceptron

4.1.1 Carving Pattern Spaces

How can researchers overcome the limitations of perceptrons? Perceptron power increases when we add one or more layers of intermediate processors are added between the input and the output units of the perceptron. These hidden units do not have any direct connection to the external world; they reside completely within the network. An artificial neural network that incorporates hidden units is called a multilayer perceptron. This chapter illustrates using a multilayer perceptron by training and interpreting the one illustrated in Figure 4-1, a network designed to detect whether an input scale is major or minor.

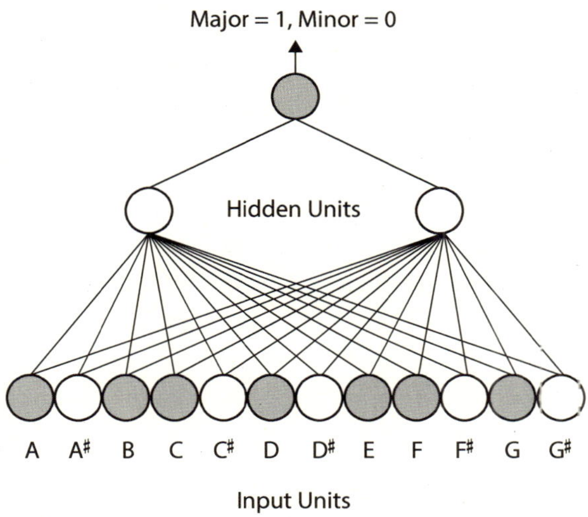

Figure 4-1 A multilayer perceptron, with two hidden units, that detects whether a presented scale is major or minor.

4.1.2 What Do Hidden Units Do?

Why do hidden units make a multilayer perceptron capable of solving problems too complex for a simpler perceptron? One answer to this question is that hidden units detect higher-order features. A higher-order feature involves relationships among different input units considered simultaneously. By including units that can detect such higher-order features, networks become capable of solving much more complicated problems (Lippmann, 1989). Indeed, adding hidden units makes artificial neural networks as powerful as any universal computing machine of interest to cognitive science (Dawson, 2013; McClelland & Rumelhart, 1986; Rumelhart & McClelland, 1986b). Thus, we should never be surprised when such a network learns to solve a musical problem. Surprises must emerge instead from our investigation of the solution that the network has discovered, and from what it might tell us about music (Dawson, 2004).

4.1.3 Training Multilayer Networks

As was the case for the perceptron introduced earlier, I will train a multilayer perceptron using a supervised learning rule that minimizes output unit error. Such learning rules modify a connection weight on the basis of two values: the activity from the unit at the input end of the connection, and the error computed for the unit at the output end of the connection (Dawson, 2004).

However, researchers encounter a problem when they attempt to define weight changes for any of the connections from an input unit in Figure 4-1 to either of the hidden units in this network. The activity at the input end of one of these connections is input unit activity. However, researchers have no idea what the error at the hidden unit end of the connection is. This is because they do not know in advance the responses that hidden units should make, and therefore they cannot define error in the same fashion as they do for the output unit.

The connectionist revolution that occurred in cognitive science in the 1980s (McClelland & Rumelhart, 1986; Rumelhart & McClelland, 1986b) began when researchers discovered a procedure for determining the error for any hidden unit in a multilayer perceptron. They found that a hidden unit's error can be specified as the sum of error signals sent to it by each of the output units to which it is connected (Rumelhart et al., 1986). Each of these signals is an output unit's error, scaled by the weight of the connection between the output and the hidden unit. In other words, output units send error signals backward through the network, and these signals determine hidden unit error, solving the credit assignment problem. Not surprisingly, the learning rule for multilayer perceptrons is often called backpropagation of error, or backprop for short. Because this rule is a generalization of the supervised learning rules for perceptrons, it is also known as the generalized delta rule.

4.1.4 Error Backpropagation

The generalized delta rule trains a multilayer perceptron to mediate a desired input/output mapping. Before training begins, a network is a "blank slate"; all of its connection weights, and all of the biases of its activation functions, start as small, random numbers. The generalized delta rule involves repeatedly presenting training patterns—input/output pairs—and then modifying weights; this was the case for perceptron training in Chapter 3.

In the generalized delta rule, a single presentation of an input/output pair proceeds as follows: The first step is to feed signals forward through the network. The multilayer perceptron's input units are then activated in order to present it an input pattern. This in turn sends signals to hidden units, which compute their net input and then their activity. Next, the hidden units send signals to the network's output units. The output units then compute their net input and activity. Output unit activities represent the network's response to the input pattern.

Second, now that output units have been activated, it is possible to measure their response error by taking the difference between the desired activity and the observed activity for each output unit. This procedure is identical to that used to train perceptrons.

Third, an output unit's error is used to modify the weights of its immediate connections. Up to this point, there is no essential difference between the supervised learning of a multilayer perceptron and the supervised learning of a perceptron.

The fourth step differentiates backprop from the training of a perceptron. In this step, each hidden unit's error is determined.This is accomplished by treating an output unit's error as if it were activity, and sending it backward as a signal through a connection to hidden units. This signal is then multiplied by the weight of the output unit's connections. Each hidden unit computes its error by summing together all of the error signals that it receives from all of the output units to which it is connected.

Fifth, once hidden unit error has been computed, the weights that feed into the hidden units can be modified using the same equation that was used to alter the weights of each of the output units. Training continues by presenting the next input/output pair in the training set, and repeating the error backpropagation procedure.

The generalized delta rule for multilayer perceptrons was initially defined for processors that use the logistic activation function (Rumelhart et al., 1986). However, variations of the algorithm exist for training multilayer perceptrons that use value units (Dawson & Schopflocher, 1992).

4.1.5 Design Decisions

One theme of this book is that artificial neural networks can only inform the cognitive science of music after their internal structure is interpreted. With this goal in mind, I gear the design decisions toward discovering the simplest network capable of solving a musical problem. A simpler network is easier to interpret than a more complex network.

My first step in achieving this goal is to determine whether the simplest network —the perceptron—can solve the problem. In Chapter 3, we found that a value unit perceptron with μs held to zero could identify scale tonics. However, at the end of that chapter we also noted that a perceptron could not learn to classify an input scale as being major or minor.

In this second case, when a perceptron is not up to the task, I next proceed to explore the more powerful multilayer perceptron. This exploration involves the same sort of design decisions used when the perceptron was investigated (learning rate, initialization, activation function, etc.). However, multilayer perceptrons require additional design decisions, such as deciding how many hidden units to include in the network.

Determining the number of hidden units requires exploring the behaviour of a number of different-sized networks. Typically, one begins with an educated guess about how many hidden units should be included. If the multilayer perceptron performs poorly on the problem (e.g., fails to learn it, stabilizes to very high error, etc.), then a different network that uses more hidden units is explored. If the network solves the problem quickly, then a network that uses fewer hidden units is trained. I may run many different simulations as I search for the simplest multilayer perceptron—that is, the one with the smallest number of hidden units—that I can reliably train to solve the problem of interest. Once I discover the simplest network for solving a musical problem of interest, I proceed to analyze its internal structure

4.2 Identifying Scale Mode

4.2.1 Task

My goal is to train an artificial neural network to distinguish major scales from harmonic minor scales—to identify scale mode. I present the network the same stimuli used to train the scale tonic perceptron in Chapter 3: either a major scale or a harmonic minor scale encoded using a pitch-class representation. I train the network to turn its single output unit "on" if the input pattern defines a major key, and to turn it "off" if the input pattern defines a minor key.

4.2.2 Network Architecture

Seeking the simplest network for accomplishing this musical task proceeds along the lines described in Section 4.1.3. As mentioned at the end of Chapter 3, we cannot successfully train a perceptron to identify key type for these different input patterns. In accordance with Section 4.1.3, the next step is to discover the simplest multilayer perceptron that is capable of identifying scale mode. After exploring a variety of different networks, the network settled upon is the multilayer perceptron illustrated in Figure 4-1.

This multilayer perceptron uses 12 input units to represent the presence or absence of pitch-classes using the same encoding described in Chapter 3 for the scale tonic perceptron. (This is because I train both networks with exactly the same set of input patterns.) The network uses a value unit as its single output unit. The output unit is trained to turn "on" if a stimulus represents a major scale, and to turn "off" if a stimulus represents a harmonic minor scale. Finally, the network uses two hidden value units as intermediate processors between its input and output units. There are no direct connections between the input units and the output unit.

4.2.3 Training Set

The training set consists of 12 different major and 12 different harmonic scales represented in pitch-class (i.e., each input stimulus was one of the rows of numbers presented in Chapter 3 as Table 3-1). As was the case for the scale tonic perceptron, each input pattern is used to turn the network's 12 different input units either "on" or "off" to indicate the presence or absence of the various pitch-classes. The desired response for an input pattern that represents a major scale is one, and the desired response for an input pattern that represents a harmonic minor scale is zero.

4.2.4 Training

The network is trained using the generalized delta rule developed for networks of value units (Dawson & Schopflocher, 1992) using the Rumelhart software program (Dawson, 2005). This program is available as freeware from the author's website http://www.bcp.psych.ualberta.ca/~mike/AlienMusic/

During a single epoch of training each pattern is presented to the network once; the order of pattern presentation is randomized before each epoch.

All connection weights in the network are set to random values between −0.1 and 0.1 before training begins. The μs of the output and hidden units are set to zero throughout training. I employ a learning rate of 0.01. Training proceeds until the network generates a "hit" for each of the patterns in the training set. I define a "hit" as activity of 0.9 or higher when the desired response is one or as activity of 0.1 or lower when the desired response is zero.

The multilayer perceptron in Figure 4-1 quickly learns to identify scale mode, typically converging after between 350 and 475 epochs of training. The network described in more detail in the next section learns to solve the problem after 390 epochs of training.

4.3 Interpreting the Scale Mode Network

How does the multilayer perceptron in Figure 4-1 detect the difference between a major scale and a harmonic minor scale, regardless of the scale's tonic? In order to answer this question let us examine the hidden unit space that confronts the output unit, as well as the input pattern features detected by the two hidden units.

4.3.1 Hidden Unit Space

The two hidden units in the network must detect stimulus properties that permit the output unit to distinguish major keys from minor keys. In order to discover what features the hidden units detect, let us plot the hidden unit space for the network. Because the multilayer perceptron uses only two hidden units, its hidden unit space can be illustrated with a two-dimensional scatterplot as is shown in Figure 4-2. In a hidden unit space, each input pattern is plotted as a point on a graph. The hidden unit activities provide the coordinates of the point. For instance, in Figure 4-2 the activity the pattern produces in Hidden Unit 1 provides a point's x-coordinate, and the activity the pattern produces in Hidden Unit 2 provides its y-coordinate. Hidden unit spaces are useful in interpreting network structure because output units can be viewed as solving a classification problem by "carving" the hidden unit space into different decision regions that separate one type of pattern (e.g., major scales in Figure 4-2) from another (e.g., minor scales in Figure 4-2) (Pao, 1989; Ripley, 1996).

A number of regularities are evident in Figure 4-2. First, all of the patterns that represent major scales fall near the origin of this space. This means that major key stimuli tend to produce near-zero activity in both hidden units. This further suggests that the role of each hidden unit is to detect (to turn on to) some property that indicates that a stimulus is not related to a major key.

Second, all of the stimuli related to major scales appear to be highly similar to one another, because all cluster closely together in Figure 4-2. In contrast, the stimuli related to minor scales spread themselves a wide distance apart. The minor scale stimuli seem to all fall roughly in a diagonal line that falls downward from left to right; there are distinct clusters of different stimuli along this line. An account of the features detected by the two hidden units should explain such regularities.

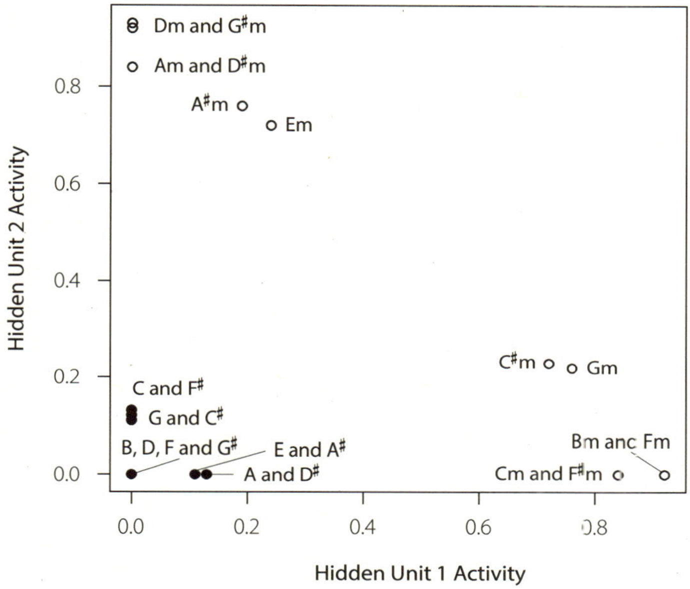

Figure 4-2 The hidden unit space for the scale mode network.

Third, in many instances different scales fall so close together in the hidden unit space that their plotted symbols fall on top of one another. For example, this is true for Dm and G♯m, and for Am and D♯m at the top left of Figure 4-2. This overlapping of symbols can make the graph harder to inspect. However, it is actually a visual property that reveals a key property of this hidden unit space: scales that seem quite different to us (like Am and D♯m) are actually nearly identical to one another as far as this network is concerned, which is why their coordinates in the hidden unit space are so close together.

4.3.2 Hidden Unit Weights

What features do the hidden units detect to determine that a stimulus represents a minor key and not a major key? In other words, what are the two hidden units detecting that permits them to position the different stimuli in the hidden unit space of Figure 4-2? In order to explore this issue let us examine the values of the weights connecting each of the input units to the hidden units.

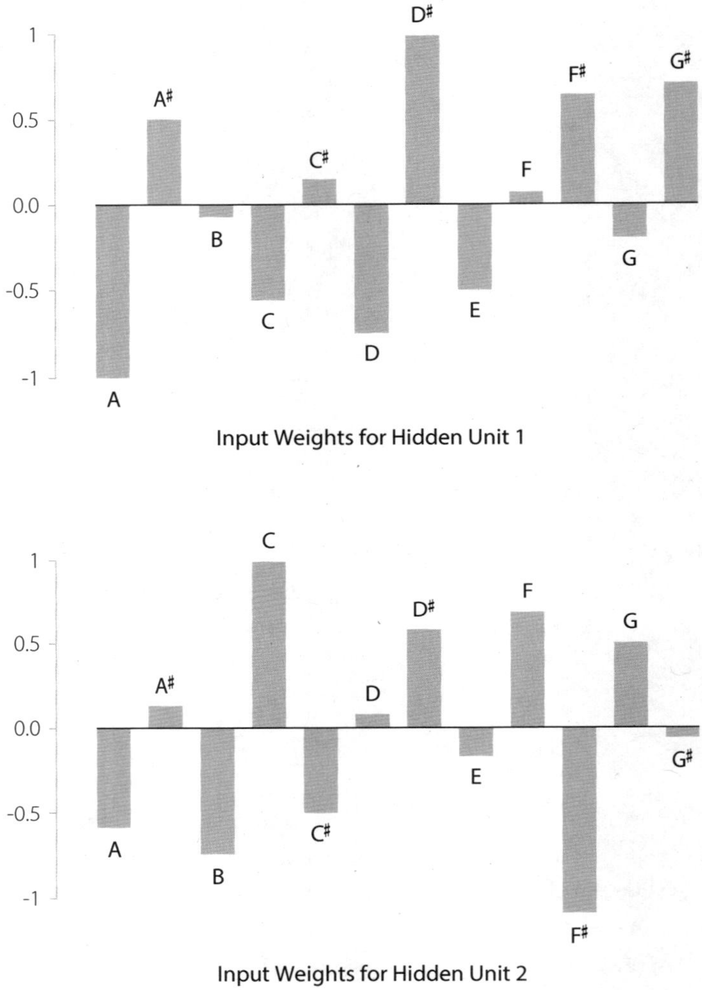

Figure 4-3 The connection weights between the 12 input units and each hidden unit in the scale mode network.

Figure 4-3 presents two bar plots, one illustrating the weights of the connections between the 12 input units and Hidden Unit 1, the other illustrating the same information for Hidden Unit 2. These two patterns of connectivity determine when particular input patterns will cause either hidden unit to generate high activity, representing the presence of a minor key (as revealed by the Figure 4-2 hidden unit space). Interpreting these connection weights should reveal what "minor scale features" are being detected by each hidden unit.

There is a high degree of regularity in both of the graphs illustrated in Figure 4-3. Exactly half of the pitch-classes in each plot have negative connection weights, and the other half have positive connection weights. Furthermore, the overall shape of each plot is identical, but the pattern for Hidden Unit 1 is "phase shifted" three pitch-classes to the left in the plot for Hidden Unit 2. Consider the pattern of bars for Hidden Unit 1 from D♯ onward to the right. The identical pattern is evident for Hidden Unit 2, but only if one begins with pitch-class C instead of D♯.

Relationships between weights in the same graph in Figure 4-3 reveal a striking property. Consider the two most extreme weights for Hidden Unit 1, the ones connecting this unit to incoming signals from A and D♯. Not only are the two weights the most extreme, but they seem almost equal in magnitude (although one weight is positive while the other is negative). These two pitch-classes are a musical interval of a tritone (six semitones) apart. If we compare other weights coming from input pitch-classes that are a tritone apart, then we see that this pattern repeats: these pairs of weights are roughly equal in magnitude but opposite in sign.

This same regularity is also evident in the Figure 4-3 plot of Hidden Unit 2 weights. For instance, the two extreme weights (from C and from F♯) are roughly equal in magnitude but point in opposite directions, and again are separated by a tritone.

Figure 4-4 presents exactly the same connection weights shown in Figure 4-3, but aligns weights from pitch-classes a tritone apart. Plotting the weights in this fashion produces two graphs in which the upper bars are a mirror image of the lower bars. This clearly indicates that weights from any two pitch-classes that are a tritone apart have the same magnitude, but are opposite in sign. This means that, for either hidden unit, if two pitch-classes a tritone apart are both turned on then their signals will cancel each other out, producing a net signal of zero.

The notion that signals from different input units cancel each other out when received by a hidden unit is of particular importance given that the multilayer perceptron in Figure 4-1 uses value units for hidden units. Recall that we define a value unit's Gaussian activation function with one parameter, μ, and that for a value unit to generate maximum activity its net input must equal μ.

In the multilayer perceptron trained to distinguish major from harmonic minor scales, the value of μ for each hidden unit, and for the output unit, is zero. Thus for either hidden unit to generate a maximum response, the incoming signal from the input units must also equal zero. This is why the notion of two signals that cancel one another is of such import.

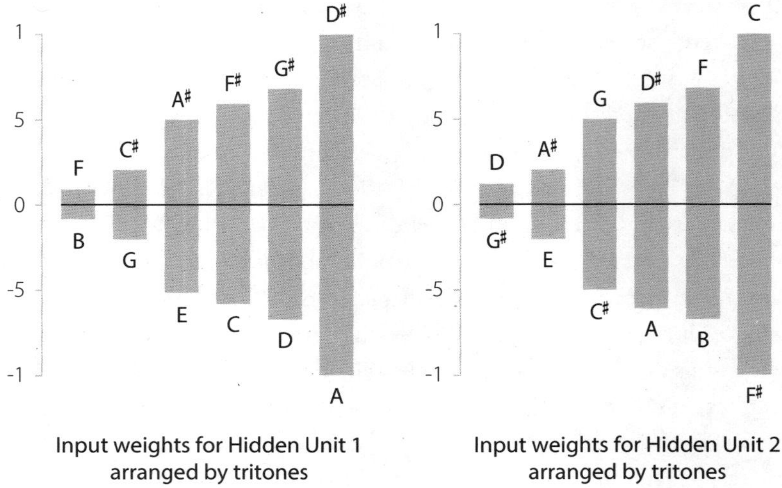

Figure 4-4 Connection weights between input units and each hidden unit.

The connection weight pattern evident in Figure 4-4 suggests that both hidden units detect tritone balance. That is, the ideal stimulus (i.e., the input pattern that produces a maximum response) is one in which pairs of pitch-classes a tritone apart are balanced. In such a stimulus, two pitch-classes a tritone apart are in the same state: they are either both off or both on. When both pitch-classes are off, their input units send a zero signal through the two connections. When both pitch-classes are on, both send signals through the two connections. However, because the weights of the two connections are equal in magnitude but opposite in sign, their signals will again cancel out, contributing little to a hidden unit's net input. Of course, it is possible for pairs of pitch-classes a tritone apart to be unbalanced. This occurs when one pitch-class is present in a stimulus but the corresponding pitch-class is not. In this situation, a definite positive or negative contribution will be added to a hidden unit's net input. This is because when the tritone is not balanced, the signals from the two pitch-classes do not cancel out. As a result, the unbalanced tritone will shift net input away from μ, making it much more likely that a hidden unit will not respond.

Figure 4-2 indicated that the two hidden units detect features true of minor keys and not true of major keys. Our analysis of connection weights indicates that this feature is "tritone balance." How do we relate this feature to the musical definition of major or minor keys?

4.4 Tritone Imbalance and Key Mode

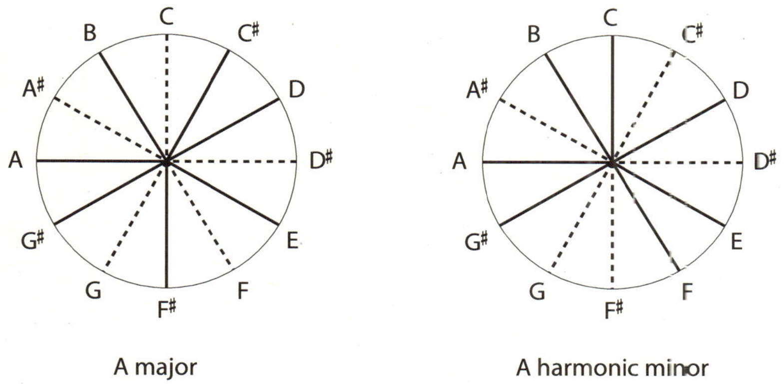

Figure 4-5 Spokes in a circle of minor seconds used to represent the pitch-classes that define major and minor keys.

In Western music, there are six possible pairs of pitch-classes separated by a tritone: [A, D♯], [A♯, E], [B, F], [C, F♯], [C♯, G], and [D, G♯]. In any stimulus presented to the network analyzed in the previous section, the more of these pairs that are balanced (i.e., their pitch-classes are both in the same state, on or off), the greater is the response of either hidden unit.

Why does the network use tritone balance to distinguish harmonic minor scales from major scales? Figure 4-5 provides an answer to this question. This figure is an alternative version of Figure 3-2, and depicts the pitch-class content of the A major and the A harmonic minor scales. Figure 4-5 differs from Figure 3-2 by indicating not only which pitch-classes are present in a scale (solid spokes) but also which pitch-classes are absent from a scale (dashed spokes).

Although Figure 4-5 provides only two example scales (A major and A minor), we saw in Chapter 3 that the two spoke patterns that it depicts can represent any major or harmonic minor scale respectively. This is because one can transpose one of the depicted scales into any other musical key by rigidly rotating the spoke pattern to a new orientation within the circle.

Figure 4-5 highlights the tritone relationships between corresponding pairs of pitch-classes. Two pitch-classes that are a tritone apart are directly opposite one another in the circle of pitch-classes. As a result, one can quickly inspect Figure 4-5 for tritone balance. If a tritone pair is balanced, then the diameter through the wheel that connects its two pitch-classes will be constant in appearance. For instance, in the A major diagram, the pair [D, G♯] balances because there is a single solid line connecting these two pitch-classes. In the A minor diagram, [D, G♯] and [B, F] are both balanced with a solid line, while [C♯, G] is balanced with a dashed line.

Recognizing that the two spoke diagrams in Figure 4-5 apply to any major or harmonic minor key respectively, two general properties are now apparent. First, major scales have almost no tritone balance. For any major scale, there will be one and only one balanced tritone. Second, for any harmonic minor scale, there will be three and only three balanced tritones. Two of these involve pairs of pitch-classes being present, while the third requires both members of a pitch-class pair to be absent. That this system requires a particular pair of pitch-classes a tritone apart to be missing is an interesting property discussed below in more detail in Section 4.5.2.

4.5 Further Network Analysis

4.5.1 Comparing Scale Geometries

Now that the hidden units of the scale mode network as tritone balance detectors have been interpreted, let us return to understanding the hidden unit space of Figure 4-2. In particular, let us explore the arrangement of the 12 harmonic minor scales in this space.

There is a long history of using spatial or geometric representations to highlight the relationships between musical entities (Hook, 2006; Krumhansl, 2005; Schoenberg, 1969; Tymoczko, 2006, 2011, 2012). When these techniques are used to map relationships between complex entities like musical scales, one feature that emerges is that major scales and minor scales tend to be mixed among one another. This is because similar scales will lie closer to one another in a musical space, and different kinds of scales can be similar to one another because they share many pitch-classes. For instance, the set of pitch-classes that defines the C major scale differs from the set that defines the A harmonic scale by only a single pitch-class. The C major scale set differs from those that define the C, D, and E harmonic minor scales by only two pitch-classes. Thus, one would expect that in a typical spatial representation C major would be near to these similar harmonic minor scales.

Figure 4-2 is interesting because it differs from this expected arrangement. Instead of surrounding minor scales with major scales as is seen in other spatial depictions (Schoenberg, 1969), the hidden unit space pulls minor scales away from the major scales at the graph's origin. Such a clear difference between the hidden unit space and other traditional geometric representations is what makes the hidden unit space interesting. A second regularity in Figure 4-2 is not only that harmonic minor scales spread along a line through the hidden unit space but also that there is a definite grouping of points along this line. In particular, points cluster together in pairs. This grouping is less evident in Figure 4-2 because in some cases paired points fall exactly in the same position. However, the pairing of points is obvious if one examines the coordinates of the minor scales in hidden unit space; Table 4-1 provides these coordinates.

Table 4-1 Properties of the 12 harmonic minor scales and their position in hidden unit space.

Scale tonic	Hidden Unit 1 activity	Hidden Unit 2 activity	Balanced tritone 1	Balanced tritone 2	Balanced tritone 3	Unbalanced pitch-classes	Hidden Unit 1 unbalanced signal	Hidden Unit 2 unbalanced signal
G#	0	0.93	[A#, E]	[C#, G]	[C, F#]	B, D#, G#	1.58	−0.16
D	0	0.92	[A#, E]	[C#, G]	[C, F#]	A, D, F	−1.59	0.16
A	0	0.84	[B, F]	[D, G#]	[C#, G]	A, C, E	−2.03	0.22
D#	0	0.84	[B, F]	[D, G#]	[C#, G]	A#, D#, F#	2.06	−0.25
A#	0.19	0.76	[C, F#]	[A, D#]	[D, G#]	A#, C#, F	0.69	0.31
E	0.24	0.72	[C, F#]	[A, D#]	[D, G#]	B, E, G	−0.70	−0.31
C#	0.72	0.23	[A, D#]	[C, F#]	[B, F]	C#, E, G#	0.29	−0.67
G	0.76	0.22	[A, D#]	[C, F#]	[B, F]	A#, D, G	−0.32	0.70
C	0.84	0	[D, G#]	[B, F]	[A#, E]	C, D#, G	0.24	2.05
F#	0.84	0	[D, G#]	[B, F]	[A#, E]	A, C#, F#	−0.24	−2.05
B	0.92	0	[C#, G]	[A#, E]	[A, D#]	B, D, F#	−0.14	−1.55
F	0.92	0	[C#, G]	[A#, E]	[A, D#]	C, F, G#	0.18	1.57

Note. Each row provides details for each scale, including the activity the scale produces in each hidden unit, the three balanced tritones, and the unbalanced pitch-classes. The final two columns provide the net input delivered to each hidden unit by the unbalanced pitch-classes.

Table 4-1 provides some additional information concerning the pairing of minor scales. It provides three columns (Balanced Tritone 1, 2, and 3) that list the three balanced tritones for each minor scale. (Note that balanced tritones 1 and 3 are pairs of pitch-classes that are both present in a stimulus, while balanced tritone 2 is a pair of pitch-classes that are both missing from the stimulus.) Examining these three columns in Table 4-1 reveals that pairs of points with identical or near-identical coordinates in Figure 4-2 represent two minor scales that have three identical balanced tritones. For instance, in the hidden unit space the inputs for the F harmonic minor scale and the B harmonic minor scale both are located at position (0.92, 0). Both of these minor scales have the same balanced tritones: [C♯, G], [A, D♯], and [A♯, E].

The row shading in Table 4-1 highlights the pairing of minor scales. If two minor scales are adjacent in the table, and have the same shading, then they are paired in the sense that they are located nearly on top of each other in Figure 4-2. Examining the scale roots (Column 1) of rows paired in this way reveals an interesting musical property that emerges from hidden unit activities: paired minor scales have roots that are a tritone apart. For instance, the harmonic minor scales for G♯ and D are both located at (0, 0.93) in the hidden unit space, and G♯ and D are a tritone apart (i.e., opposite one another in Figure 4-6).

The proximity relations among minor scales in the hidden unit space are based on a definite musical property (shared balanced tritones), and produce a very regular pairing of scales. However, the balanced tritones property is an atypical musical regularity in comparison to more traditional spatial representations. As a result, the proximity relations in the hidden unit space are markedly different from proximity relations in more traditional spaces.

Consider the following set of harmonic minor scales: G♯m, Dm, Bm, and Fm. Traditional spatial representations would plot G♯m closer to Bm and Fm than to Dm. This is because it differs from the former two scales by only two pitch-classes but differs from the last one by three. Another way to consider the similarity is that the tonic (G♯) of G♯m is a minor third or three semitones away from the tonic of either Bm or Fm. However, the tonic of G♯m is a full tritone or six semitones away from the tonic of Dm. In a traditional spatial depiction of these four scales or keys, Dm is farther away from the other three.

Figure 4-6 provides two examples of spatial arrangements based on measures from traditional music theory. Multidimensional scaling (MDS) was used to analyze distances between scales (where distance is a function of the number of pitch-classes shared by two scales) into a spatial map in which scales that are similar to one another are located closer together. The left part of Figure 4-6 provides the two-dimensional MDS solution; it accounts for 58.1% of the variance in the original distance matrix. It places all of the major scales in a well-known musical pattern: the circle of perfect fifths. It also arranges the minor scales around a separate circle of perfect fifths, and places this circle inside the circle of major scales. The two different circles of perfect fifths in this MDS solution have different orientations; for instance, the inner circle of harmonic minor scales is oriented so that the two scales closest to A major are F♯ minor and B minor. Note that within this MDS solution G♯m is closer to Bm and Fm than it is to Dm.

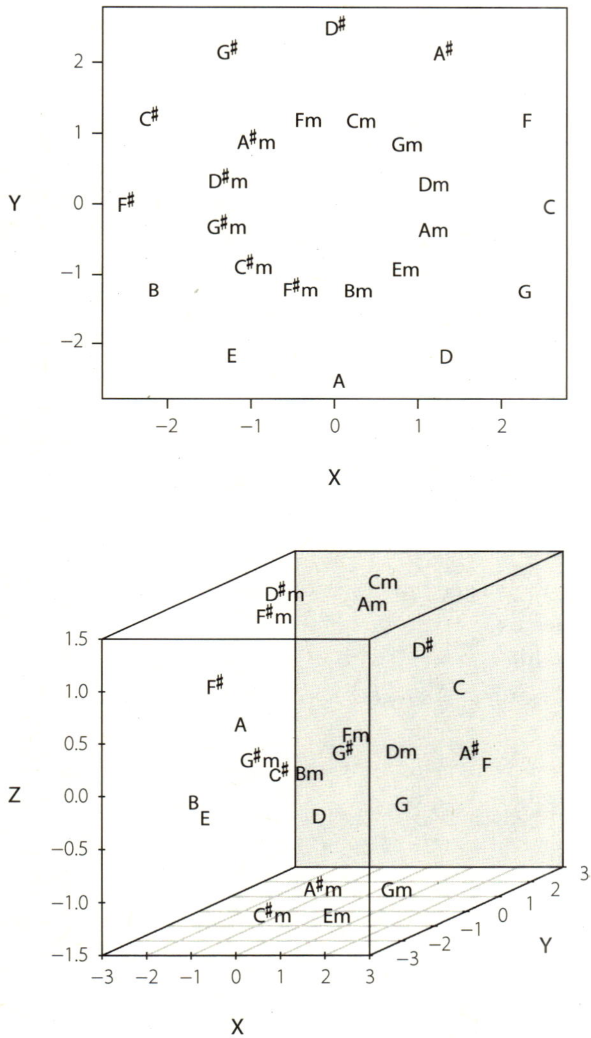

Figure 4-6 A two-dimensional and a three-dimensional multidimensional scaling solution that arranges scales associated with different keys in a spatial map.

The right side of Figure 4-6 presents the three-dimensional MDS solution for the scale distances matrix. The added dimension improves the fit to the data; it accounts for 69.2% of the variance in the original distance matrix. The third dimension pulls different sets of minor scales away from one another in the vertical direction. At the top of the cube one finds D♯m, Cm, Am, and F♯m; in the middle of the cube G♯m, Fm, Dm, and Bm; at the bottom of the cube C♯m, G♯m, Gm, and Em. Once again, G♯m is closer to Bm and Fm than to Dm. There is also some vertical separation between major scales in this solution, but all are generally positioned outside and around the middle of the cube. What is the relationship between the two MDS solutions in Figure 4-6? It appears that if one were to project the positions of the scales in three-dimensional MDS solution down to the bottom plane of the cube, the result would be the two circles of perfect fifths that are apparent on the left of Figure 4-6.

Now let us compare the spatial arrangements of scales based on traditional theory (Figure 4-6) with the spatial arrangements of scales in the hidden unit space (Figure 4-2). In particular, let us continue to focus upon the four minor scales G♯m, Bm, Fm, and Dm. The spatial relationships between these four harmonic minor scales are quite different in the hidden unit space. First, rather than being farther apart, scales whose roots are a tritone apart (G♯m and Dm, or Bm and Fm) have nearly identical locations in Figure 4-2. Second, rather than being closest to scales a minor third away, scales that differ by this amount are very far apart in the hidden unit space. In particular, the points for G♯m and Dm are the two that are farthest away from the points for Fm and Bm in Figure 4-2. Table 4-1 shows the coordinates for the latter two scales are the reflection of the coordinates of the former [(0.92, 0) vs. (0, 0.92)]. Why are different scales with similar balanced tritone structure so far apart in Figure 4-2? The three pitch-classes that are not part of a balanced tritone must be responsible.

When we present a harmonic minor scale to our multilayer perceptron, the three unbalanced pitch-classes are in essence the only source of net input to either hidden unit. This is because all other pitch-classes are balanced, and therefore have near zero to net input. The final three columns of Table 4-1 provide information regarding the effects of unbalanced pitch-classes on a scale's position in the hidden unit space.

The first of these three columns simply lists, for each harmonic minor scale, the three pitch-classes that are unbalanced. The remaining two columns provide the net input to each hidden unit that is only due to the three unbalanced pitch-classes. These columns reveal some interesting properties concerning the spatial arrangement of the hidden unit space.

First, consider two scales that have identical balanced tritones (e.g., Fm and Bm). These two scales differ from one another in terms of their three unbalanced pitch-classes. For Fm these are F, G♯, and C, and for Bm these are B, D, and F♯. Look in Table 4-1 at the net inputs that each of these unbalanced sets produces in each hidden unit. One of these sets produces net inputs that are essentially equal in magnitude, but opposite in sign, to the net inputs produced by the other set. Remember that the symmetric form of the Gaussian activation function means that it in essence ignores the sign of net inputs. Thus while these two different sets of unbalanced pitch-classes send different net inputs to the hidden units, the hidden units generate the same activity to either set, so the two patterns wind up in the same position in the hidden unit space. This account holds for any of the paired scales in Table 4-1

Now consider a different pair of scales, G♯m and Dm, whose balanced tritone structure is similar to that of Fm and Bm. The unbalanced pitch-classes for G♯m and Dm produce the same magnitude of net input (ignoring sign) to the hidden units as do the unbalanced pitch-classes for Fm and Bm. However, they send this magnitude to the opposite hidden units! As a result, the position of these two scales is reflected in the hidden unit space (i.e., the coordinates [x, y] of Fm and Bm become the coordinates [y, x] of G♯m and Dm).

This analysis leads to two general musical statements about the positioning of harmonic minor scales along the line through hidden unit space. First, two scales whose roots are a tritone apart will have the same position on this line. Second, two scales whose roots are a minor third apart will fall at positions reflected across the centre of this line.

The discussion of the geometry of Figure 4-2 in this section indicates that while the proximity relations in Figure 4-2 (made clear in the Table 4-1 coordinates) are quite different than traditional ones (e.g., Figure 4-6), they are based upon musical structure. The atypical arrangement of scales in the hidden unit space is a consequence of the multilayer perceptron detecting a particular musical feature, balanced tritones. Because of this musical regularity, scales that typically are viewed as being distantly related become highly similar, and scales that are typically viewed as being highly similar become distantly related. In short, paying attention to the hidden unit space can reveal novel properties that are still completely consistent with formal music theory.

4.5.2 The Missing Balanced Tritone

The hidden units of the scale mode network detect tritone balance. This property, reflected in the connection weights from input units to hidden units (Figure 4-4), recognizes that a harmonic minor scale contains three sets of balanced tritones, two of which are present in the scale; the third pair is balanced because its two pitch-classes are both absent. This balance is critical for the function of the hidden units: balanced tritones contribute little to hidden unit net input, and therefore cause high hidden unit activity because the μ of each Gaussian activation function is zero. This in turn causes the arrangement of minor scales in the Figure 4-2 hidden unit space.

The tritone that balances because both of its pitch-classes are absent is critical to the ability of hidden units to separate harmonic minor scales from major scales in the hidden unit space. It also provides an interesting twist on music theory. This is because the necessary absence of tritones is not a traditional component of music theory. For instance, consider using mathematical set theory to explore musical formalisms (Forte, 1973; Roig-Francolí, 2008; Straus, 2005). One contribution of set theory is the ability to generate, for some set of pitch-classes, a six-dimensional vector called an interval class vector or an ic vector. An ic vector provides information about the frequency of occurrence of different musical intervals in a musical entity, where a musical interval is the distance between two pitch-classes measured in semitones. In Forte's system, the set of pitch-classes that define any harmonic minor scale produces the ic vector 335442. The last digit in this ic vector indicates the presence of two tritone intervals (i.e., in terms of Figure 4-5 two balanced pairs of tones present in the scale). Similarly, the ic vector for any of the major scale stimuli presented to the scale mode network is 253461. Its last digit indicates that a major scale includes only a single pair of pitch-classes a tritone apart, as was earlier illustrated in Figure 4-5. In other words, ic vectors make explicit the well-known property that a major scale includes only a single tritone interval. It might be tempting to conclude that the scale mode perceptron simply uses the number of present tritones to distinguish the major mode from the harmonic minor mode.

Importantly, what the ic vector for a harmonic minor scale fails to make explicit is the absence of the two additional pitch-classes that represent the third balanced tritone (e.g., the absence of both C♯ and G in the A harmonic minor diagram in Figure 4-6). Identifying which pitch-classes are absent from a musical object might be an odd way for a human musical analyst to think about musical scales, but it arises naturally in this artificial neural network. Indeed, when the nature of the absent balanced tritone is considered, we encounter some interesting music theory insights.

In Section 4.5.1, we noted that the arrangement of minor scales in Figure 4-2 was affected by the three pitch-classes that are not part of a balanced tritone. For each harmonic minor scale, Table 4-1 provides the three unbalanced pitch-classes in its final three columns. From the perspective of music theory, the unbalanced pitch-classes suggest an elegant and simple musical interpretation: they are the three pitch-classes that define the minor triad of the harmonic scale. This means that for the minor stimuli, Figure 4-2 also provides a spatial arrangement not of harmonic minor scales but instead of minor triads.

Other spatial representations have been developed to represent the relationships between triads or chords (Hook, 2006). In Hook's Tonnetz, each node represents a triad; connections between nodes represent a possible progression from one triad to another. Furthermore, each connection between nodes represents an efficient voice leading. This is because linked triads share two pitch-classes, even though they are opposite in mode (i.e., minor vs. major). In other words, Hook's (2006) representation is similar to the scale relations that we considered earlier to create Figure 4-6.

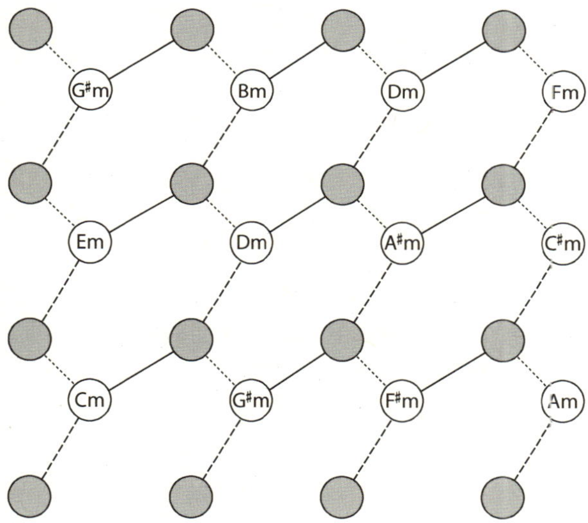

Figure 4-7 The structure of Hook's (2006) Tonnetz for triads.

The nearest neighbours of triads aligned in the same row in Figure 4-7 are related by a short musical interval (a minor third); the nearest neighbours of Bm are G♯m and Dm. Furthermore, triads related by a longer musical interval are not nearest neighbours. For instance, Bm and Fm are farther apart in Figure 4-7.

We are now in a position to contrast the spatial arrangement of triads in Figure 4-7 with the hidden unit space of Figure 4-2. It was noted above that a consequence of the missing balanced tritone is that the positions of harmonic minor scales in the hidden unit space also provide the positions of minor triads. The proximity relations among minor triads in the hidden space are quite different from those in the triad Tonnetz. In particular, minor triads that are farther apart in Figure 4-7 occupy identical locations in the hidden unit space (e.g., Bm and Fm). In addition, nearest neighbours in the triad Tonnetz are quite far apart in the hidden unit space. For instance, G♯m and Dm, the two triads closest to Bm in Hook's space, are the farthest away from Bm in the hidden unit space. Of course, the fact that the hidden unit space arranges minor triads in a line, while Hook's Tonnetz arranges them in a grid, is another crucial difference between the two. None of these differences between the two spaces should be particularly surprising: hidden unit space locations reflect tritone structure, while Hook's (2006) triad Tonnetz bases proximity in terms of shared pitch-classes.

If we interpret the hidden unit space as establishing certain proximity relations among minor triads, and these differ from those in traditional geometric models of harmonic progression or voice leading, then what does the hidden unit space imply about chord progressions? The key feature of the hidden unit space is that two minor triads that are a tritone apart have the same location, and therefore are equivalent. In terms of progressions, this suggests that one can easily change from one triad to another that is a tritone away. Interestingly, this implication of the hidden unit space parallels the tritone's relevance to jazz.

One common technique to add variety to jazz chord progressions is the use of chord substitutions. In this practice, one chord in the progression is replaced with another musically related chord. For example, in tritone substitution a dominant seventh chord in one key is replaced with the dominant seventh chord from a key that is a tritone away. This is possible because both dominant seventh chords contain the same two notes a tritone apart, making the original changes harmonically similar to the changes created by the tritone substitution (Tymoczko, 2008).

Tritone substitution is possible (i.e., sounds musically correct) in jazz because the two dominant seventh chords contain exactly the same tritone, and the two keys a tritone apart share enough pitch-classes to make them harmonically similar (Tymoczko, 2008). In certain respects, the multilayer perceptron is detecting analogous structure in harmonic minor scales, organizing them in such a way that the nearest neighbour to a minor scale in the hidden unit space will be a scale with identical underlying tritone structure. Interpreting these scale positions as being the positions of minor triads reveals that the hidden unit space of the multilayer perceptron may have a special affinity for tritone substitution.

4.6 Summary and Implications

This chapter explored the properties of a multilayer perceptron trained to identify the mode of a stimulus scale (major or harmonic minor). A network that employed two hidden value units accomplished this task. Further investigation of the internal structure of this network revealed that both hidden units detected a particular property: tritone balance. The network learned that harmonic minor scales have three balanced tritones while a major scale has only one.

The scale mode network's musical theory departs from traditional theory. In particular, the functioning of this network depends upon an absent balanced tritone: two pitch-classes, separated by a tritone, which are both missing from a harmonic minor scale. Furthermore, by considering this tritone in combination with the two present balanced tritones we discovered that the only unbalanced pitch-classes in a harmonic minor scale define a minor triad. This observation is a new addition to music theory.

The most telling indication that the network's emphasis on balanced tritones represents a departure from traditional music theory is to consider that it leads to a spatial arrangement of scales (and of minor triads) that is markedly different from those based on traditional notions of musical relationships like shared pitch-classes. Figures 4-6 and 4-7 arrange musical entities in a map whose distances reflect the number of shared pitch-classes. Both of these maps interweave major and minor musical elements. In contrast, the spatial arrangement the network provides—its hidden unit space illustrated in Figure 4-2—is markedly different. Major and minor musical elements do not interweave; instead, a linear arrangement of minor harmonic scales reveals distance relationships based on shared tritone structure instead of shared pitch-classes.

The novel spatial arrangement of scales that the network reveals provides some interesting suggestions regarding composing music. One fundamental principle of composition is modulation, in which a rational structure is used to change from one musical key to another midway through a piece. Schoenberg's (1969) spatial map of keys provides a spatial guide to such modulation. One can modulate from the current musical key to another nearby in the map without musical disruption because keys that are near one another in the map have shared properties (e.g., they have many pitch-classes in common).

The proximity relationships between minor scales in the hidden unit space of Figure 4-2 space suggest an alternative approach to explore for modulating between minor keys. In particular, the hidden unit space suggests that one may be able to modulate between one minor key and another that is a full tritone away, not because of shared pitch-classes but instead because of shared balanced tritones. Furthermore, the hidden unit space suggests the possibility of successful modulation to keys other distances away, because these scales are close to one another in the hidden unit space. For instance, G♯m and Dm are closest to D♯m and Am in the space. Modulating from G♯m to D♯m is part of common practice (these two keys are a perfect fifth apart), as does modulating from Dm to Am (these two keys are a perfect fourth apart). However, the same geometry suggests less typical modulations a minor second apart: from G♯m to Am, or from Dm to D♯m. The hidden unit space, like more traditional spatial representations, is a source of alternative compositional ideas.

5

Networks for Key-finding

5.1 Key-Finding

5.1.1 Key-Finding Outlined

As was noted earlier in Chapter 1, one of the most important discoveries about music cognition is that listeners mentally represent music using a tonal hierarchy (Krumhansl, 1979, 1990a; Krumhansl, Bharucha, & Kessler, 1982; Krumhansl & Shepard, 1979). The tonal hierarchy organizes the pitches of a particular key to reflect the fact that some are more important than others are, and that other tones must be considered in relation to these stable pitches. For instance, if a musical context establishes a specific musical key, then the best-fitting note is the tonic of that key (i.e., the pitch in position 1 of the scale). For example, in the musical key of C major the note C is the most stable. The next best tones are those in either the 3 or 5 positions of the key's scale. In the key of C major, these are the notes E or G. Less apt than these two notes are any of the set of remaining notes that belong to the scale. In the context of C major, these are the notes D, F, A, or B. Finally, the least stable tones are the set of five pitch-classes that do not belong to the context's scale. For C major, these are C♯, D♯, F♯, G♯, and A♯.

Krumhansl (1990a) employs the tonal hierarchy as the basis for a cognitive theory of music perception. Her central claim is that music perception does not simply depend upon music's acoustic properties but also depends upon the mental organization of musical sounds. Listeners have precise knowledge about the structure of tonal music, and use this knowledge to understand music. One consequence of Krumhansl's theory is that listeners must possess a fundamental ability to identify the tonal centre of music. If one cannot identify a musical key, then one cannot use tonal hierarchies to organize music. Human listeners are indeed able to infer musical key on the basis of very little musical evidence (Butler, 1989).

Given the importance of the tonal hierarchy in musical cognition, and the consequent importance of identifying tonal centres, it is not surprising that many researchers propose theories about how listeners identify musical keys. We call these theories about key-finding or tonal implication. When one hears a musical selection in the key of C major, what procedure does one use to infer that the piece is indeed in that key, to permit using the C major tonal hierarchy to understand the selection's tonal structure? A key-finding theory aims to answer this question.

One influential theory of key-finding is proposed by Krumhansl and Schmuckler, and described in detail in Krumhansl's seminal work *Cognitive Foundations of Musical Pitch* (Krumhansl, 1990a). This algorithm, described in more detail later in this chapter, compares a musical input to templates associated with different musical keys. These templates are derived from the probe-tone method. The template that provides the best match to the musical input is used to assign a key to the input. Variations of this algorithm have also been proposed; they maintain its general character but include technical modifications to improve performance even further (Albrecht & Shanahan, 2013; Shmulevich & Yli-Harja, 2000; Temperley, 1999, 2007). The model also has status as a cognitive theory of key-finding with support from psychological experiments (Frankland & Cohen, 1996; Schmuckler & Tomovski, 2005).

Other key-finding theories also exist. Some propose that listeners detect the presence of rare musical intervals like the tritone (Brown & Butler, 1981; Browne, 1981; Butler, 1989). Others describe algorithms that match musical inputs to particular musical patterns, such as specific key-implying pitch sequences (Handelman & Sigler, 2013; Holtzman, 1977), positions in a geometric space defined by pitch-class coordinates (Albrecht & Shanahan, 2013), or pitch-classes that belong to particular major or minor scales (Longuet-Higgins & Steedman, 1971; Vos & Van Geenen, 1996). This chapter explores still other approaches to key-finding: algorithms based upon artificial neural networks.

5.1.2 One Network for Key-Finding

A key-finding procedure must accomplish two different tasks. First, when presented a musical stimulus, it must judge the tonic of the stimulus's musical key. Second, it must judge the mode (major vs. minor) of the key. The perceptron described in Chapter 3 identifies the tonic of a presented musical scale but does not identify mode. The multilayer perceptron described in Chapter 4 identifies the mode of a presented musical scale but does not identify tonic. Thus, neither network is a candidate for a key-finding algorithm.

One could merge these two networks together, unaltered, to create a system capable of identifying both tonic and mode. This merging is possible because both networks use identical input representations and are trained on identical patterns; but the tonic judgment of the perceptron is accomplished by a set of connection weights and output units that are completely independent of the connection weights, hidden units, and output unit of the multilayer perceptron.

However, it is not typical for connectionist researchers to train different networks on task components, and then sew them together into a working whole like Frankenstein's monster. Instead, a more typical approach is to train a single artificial neural network to accomplish the entire task at once. This chapter explores two such networks for key-finding.

5.2 Key-Finding with Multilayered Perceptrons

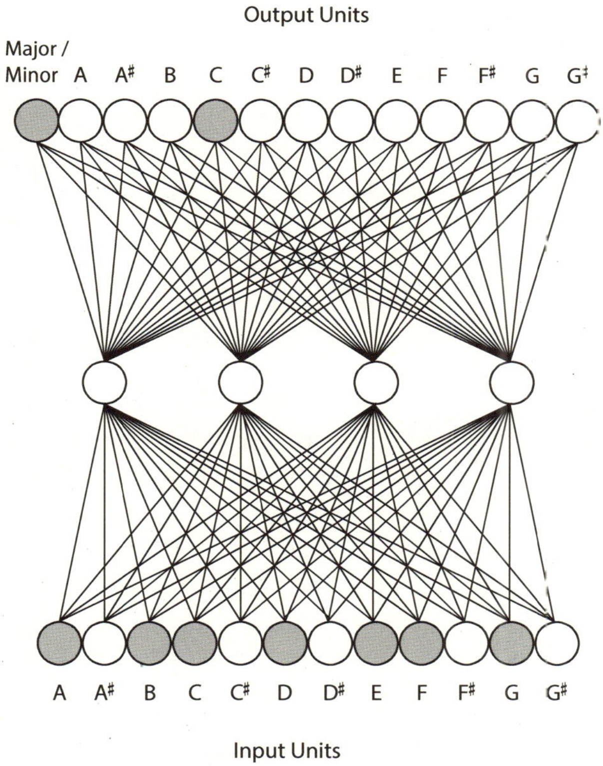

Figure 5-1 A multilayer perceptron that uses four hidden units to detect both the mode and the tonic of presented major or harmonic minor scales.

To begin our exploration of key-finding with artificial neural networks, let us continue to train a network to generate responses when using musical scales as stimuli (i.e., the sets of pitch-classes from Table 3-1). In Chapter 3, a perceptron learned to output the tonic for each of these stimuli, while in Chapter 4 a multilayer perceptron learned to output the mode. The first network to be described in the current chapter is a multilayer perceptron (Figure 5-1) that learns to respond with both the mode and the tonic of each of these scales.

5.2.1 Task

The goal is to train a single artificial neural network to identify both the tonic and the mode when presented either a major scale or a harmonic minor scale. Given a main result of Chapter 3—that a multilayer perceptron was required to detect scale mode—the network of interest in this section uses a layer of hidden units to solve this problem.

5.2.2 Network Architecture

Figure 5-1 illustrates a multilayer perceptron for the scale-based version of the key-finding problem. It uses 13 output units to represent scale mode and tonic and uses 12 input units to represent the pitch-classes that belong either to a major or to a harmonic minor scale. Pilot simulations revealed that this network requires four hidden units to solve this problem. All of these hidden units, and all of the output units, are value units.

5.2.3 Training Set

The training set consists of 12 different major and 12 different harmonic scales represented in pitch-class format (i.e., each input stimulus is one of the rows of numbers presented in Chapter 3 as Table 3-1). The representation of inputs is identical to that described earlier in Chapters 3 and 4. The representation of outputs uses 13 value units to combine the representations used by the Chapter 3 perceptron and by the Chapter 4 multilayer perceptron. One output unit is trained to turn on if a presented scale is major, and to turn off otherwise. The remaining output units each represent a possible scale tonic; the network is trained to turn the unit representing the correct tonic on and to turn the other tonic units off.

5.2.4 Training

The network is trained with the generalized delta rule developed for networks of value units (Dawson & Schopflocher, 1992) using the Rumelhart software program

(Dawson, 2005). All connection weights in the network are set to random values between −0.1 and 0.1 before training begins. The μs of the output and hidden units are all initialized to zero, but are modified as training proceeds. A learning rate of 0.01 is employed. Training occurs until the network generates a "hit" for every output unit for each of the patterns in the training set. Again, a "hit" is defined as activity of 0.9 or higher when the desired response is one or as activity of 0.1 or lower when the desired response is zero. The multilayer perceptron described in more detail in the next section learns to solve the problem after 6801 epochs of training.

5.3 Interpreting the Network

5.3.1 Carving Hidden Unit Space

In general, multilayer perceptrons solve classification problems by using output units to carve a hidden unit space into various decision regions. All of the output units in the Figure 5-1 network are value units. A value unit carves a hidden unit space into three areas using two parallel straight cuts. Patterns that fall between the two cuts (which are very close together) turn the unit "on," and patterns that fall outside the cuts turn the unit "off." In order for the multilayer perceptron to solve the key-finding problem, its hidden units must arrange patterns in a hidden unit space so that the output unit for mode can separate any major scale from all of the harmonic minor scales. This arrangement must also permit a tonic output unit to separate the two scales that have its tonic as a root from all of the other scales.

It is easy to imagine hypothetical hidden unit spaces that permit the output units to operate in this fashion. Figure 5-2 illustrates a hypothetical two-dimensional space. It arranges the scales in order by root, so that scales that have the same root are side by side. It also arranges the scales by type, so that all the major scales line up on the left, and all the harmonic minor scales line up on the right. As illustrated in Figure 5-2, this arrangement can easily be carved to identify C major by having the mode output unit "carve out" the major scales vertically, and by having the tonic output unit for C carve the two C scales out horizontally.

In order to solve the problem with the hypothetical space illustrated in Figure 5-2, network training must accomplish two things. First, the connection weights between the input units and the hidden units have to take on values that permit the hidden units to position the scales in appropriate locations in the space. Second, the connection weights between the hidden units and the output units have to take on values that permit the output units to carve the space appropriately.

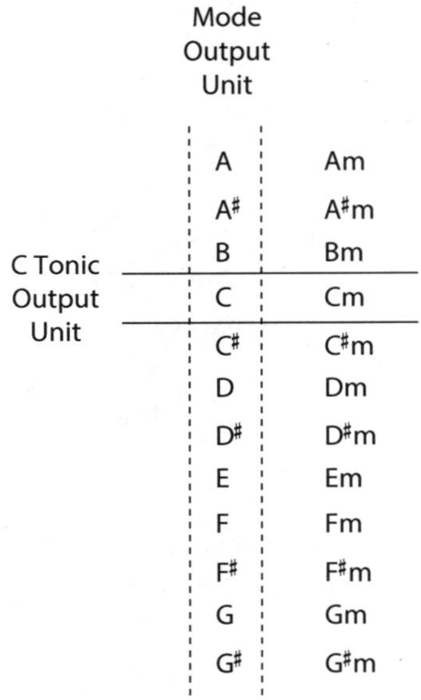

Figure 5-2 A hypothetical two-dimensional hidden unit space for key-finding.

A two-dimensional space (e.g., the one illustrated in Figure 5-2) does not provide the only possible solution for the key-finding problem. The fact that four hidden value units are required to permit the multilayer perceptron of Figure 5-1 to converge to a solution indicates that it needs a four-dimensional hidden unit space to classify the different scales. Let us take a moment to consider the properties of the four-dimensional hidden unit space obtained from the trained multilayer perceptron. This hidden unit space contains 24 different points, one for each stimulus scale. The four-dimensional coordinates of each point are the four activities produced in the network's hidden units by a particular stimulus.

The problem with a four-dimensional hidden unit space is that it is impossible to visualize. However, we can understand a portion of its structure by considering lower-dimensional visualizations. For example, Figure 5-3 presents a three-dimensional visualization of part of this hidden unit space by plotting the positions of scales in a cube using activities of Hidden Units 1, 2, and 4 to provide the coordinates of each scale in the space.

In the discussion of Figure 5-2, I observed that it was necessary to arrange the major scales so that they are separable from all of the harmonic minor scales. Figure 5-3 indicates that the hidden unit space for the multilayer perceptron achieves this goal. Note that all of the major scales arrange themselves in a group near the front of the cube. The dashed lines suggest where this cube could be carved to separate these scales, which turn the output unit on, from all of the harmonic minor scales that are spread elsewhere in the space and which turn the unit off.

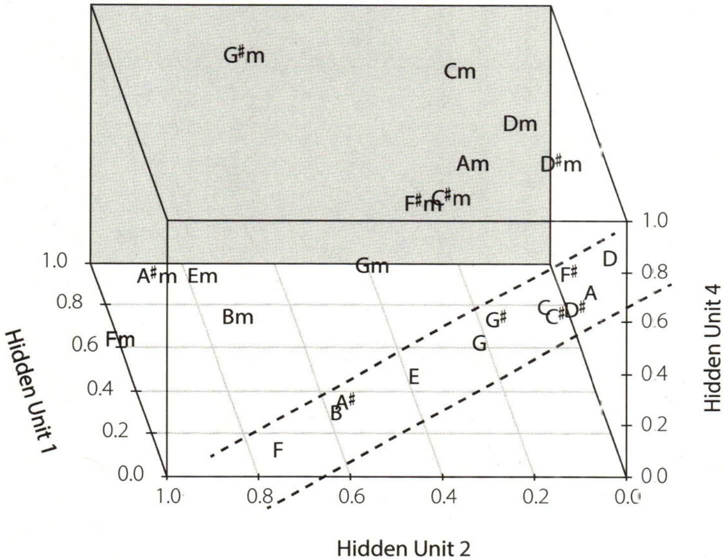

Figure 5-3 A three-dimensional projection of the four-dimensional hidden unit space for the key-finding multilayer perceptron.

In the earlier discussion of Figure 5-2, it was noted that a major scale and a harmonic minor scale with the same root must be positioned in the hidden unit space so that a single unit (the output unit for their tonic) could separate both from all of the other scales. Figure 5-4 presents a second three-dimensional projection of the four-dimensional hidden unit space that attempts to show how it arranges different scales based upon the same root. It uses Hidden Unit 1, 2, and 3 activities to provide the coordinates of each scale in the cube. An examination of Figure 5-4 indicates that different scales with the same tonic are aligned in the cube in such a way that they can be separated from the others by the two parallel cuts of a value unit. The dashed lines in Figure 5-4 provide a sense of how this might be accomplished for the tonic F♯. Importantly, there is no requirement that one tonic output unit in the multilayer perceptron carve the hidden unit space in a fashion that is related to the carving made by other output units (e.g., making cuts at different positions but in the same direction). Thus from the perspective illustrated in Figure 5-4 there is no compelling arrangement of all of the scales (e.g., around a circle of minor seconds). All that is required in this space is that one hidden unit can separate two locations (the locations of the major and minor scale with the same tonic) from all of the others.

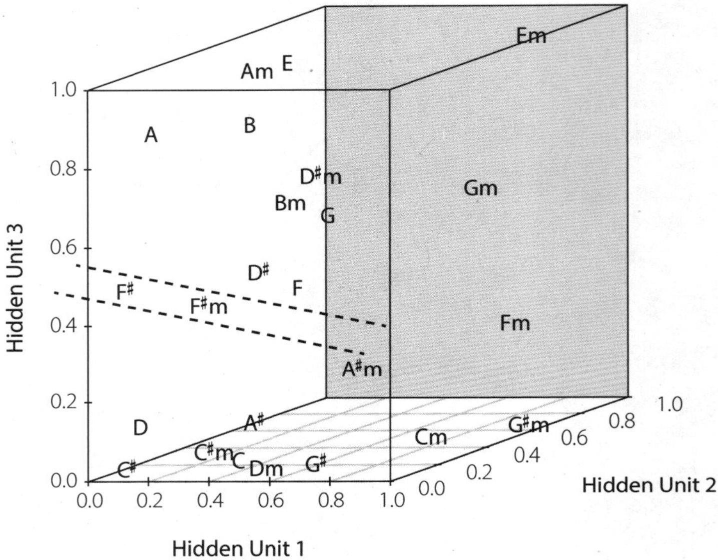

Figure 5-4 A different three-dimensional projection of the four-dimensional hidden unit space for the key-finding multilayer perceptron.

5.4 Coarse Codes for Key-Finding

The foregoing discussion emphasized hidden unit space regularities. In particular, it focused on the arrangement of the different scales in a four-dimensional hidden unit space in order to permit output units to isolate particular scales for generating correct responses for the key-finding task. We now turn to a related topic: What musical features does each of the four hidden units detect?

Some of the seminal advances in neuroscience were made possible by presenting stimuli to the sensory systems of animals while recording responses from individual neurons (Hubel & Wiesel, 1959; Lettvin, Maturana, McCulloch, & Pitts, 1959). With this technique, it is possible to describe a neuron as being sensitive to a trigger feature, a particular pattern which, when presented, produces maximum activity in the cell. This success, in turn, led some researchers to propose a neuron doctrine for research on perception (Barlow, 1972, 1995). Adherents to the neuron doctrine believed that a complete theory of perception would result from knowing the trigger features of each perceptual neuron. One can attempt to apply the neuron doctrine to the interpretation of artificial neural networks by identifying those stimuli that produce maximum activity in each of a network's hidden units (Dawson, 2004, 2013).

In order to explore the trigger features of the four hidden units of Figure 5-1, one can wiretap them by recording the activity produced in each hidden unit by each pattern in the training set. Given the nature of the key-finding problem, one might expect that some hidden units are dedicated to detecting properties related to scale tonics, while others specialize in detecting properties related to scale modes.

Figure 5-5 presents the wiretapping results. Each panel of the figure presents the results for one hidden unit. The length of each horizontal bar provides the activity produced by one stimulus. The stimuli are grouped by tonic, with the major scale on the left and the minor scale on the right. Note that activity ranges from zero to one in both directions across the figure.

Figure 5-5 indicates that the hypothesis that the four different hidden units are specialized is mistaken. There is no indication that some hidden units detect scale mode and that others detect scale tonic. Instead, all four hidden units respond to a subset of tonics, and generate strong responses to a subset of scales that includes both majors and harmonic minors. In other words, the wiretapping does not reveal any illuminating trigger features.

Consider Hidden Unit 1, the upper left panel of Figure 5-5. It prefers harmonic minor scales, because there seems to be more black displayed by the bars on the right than there is grey displayed by bars on the left. However, it does not respond to every minor scale: it produces no activity to Bm, and produces minimal activity to A♯m, C♯m, and F♯m. Furthermore, it generates strong activity to some of the major scales (D♯, G, and G♯). In short, if this hidden unit is a scale mode specialist, then it is not a very accurate one.

It is also the case that Hidden Unit 1 does not appear to be successful in the specialized role of tonic detector. It generates strong responses to four tonics (C, D♯, G, and G♯) regardless of scale mode and generates very weak responses to several other tonics (B, C♯, and F♯). In some instances (F) it responds to the major scale with this tonic, but hardly responds at all to the harmonic minor scale with the same tonic.

An examination of Figure 5-5 indicates that similar stories are true of the other three hidden units as well. They tend to have a general preference for one scale mode or the other, but can generate strong responses to either. They tend to have high activations to some tonics and not others, but sometimes activate to a tonic in one scale mode but fail to activate to the same tonic in a different scale mode. Individually, none of the hidden units seems eminently useful for solving the key-finding problem. However, collectively they are all important, because the multilayer perceptron uses these hidden units to converge to a solution. How is it possible for hidden units to be individually poor at representing a problem's solution, but collectively successful?

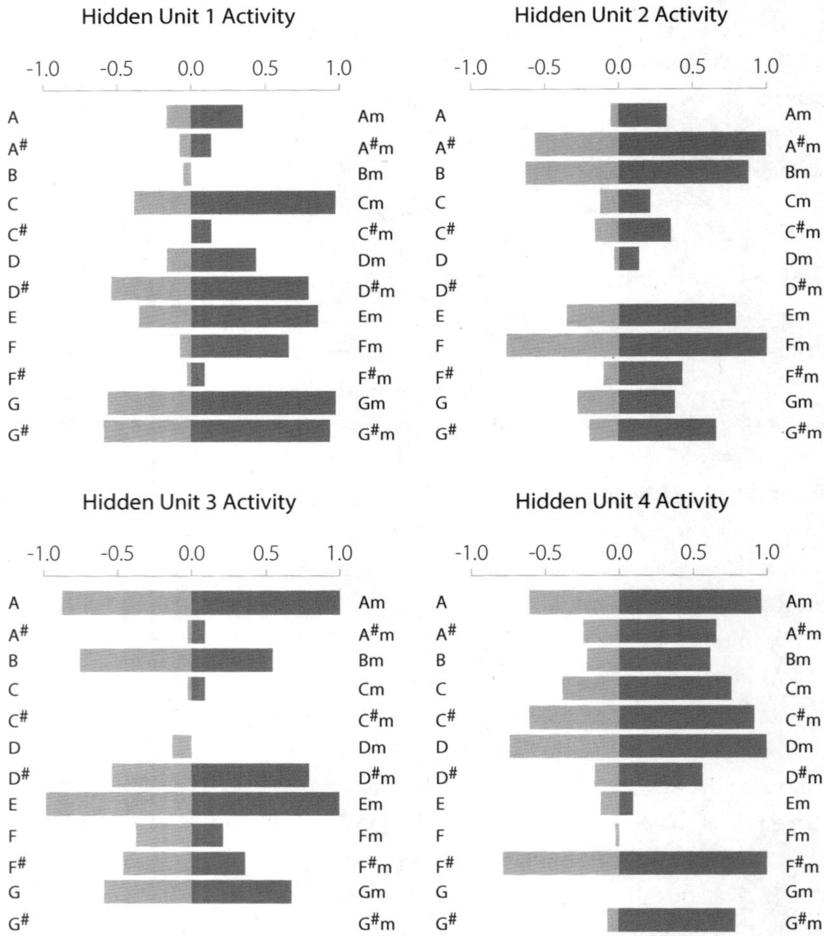

Figure 5-5 The results of wiretapping each hidden unit, using the 24 input patterns as stimuli.

Figure 5-5 suggests one answer to this question. Let us consider the four hidden units in terms of tonic identification. Imagine a scale stimulus that produces little activity in Hidden Unit 1, moderate activity in Hidden Units 2 and 3, and high activity in Hidden Units 4. What is the tonic and mode of this scale? For Hidden Unit 1, several scales produce activity of 0.1 or lower, such as: C♯, Bm, F♯, B, F, A♯, and F♯m. For Hidden Unit 2, a different subset of stimuli generates moderate activity between 0.29 and 0.61: Am, C♯m, Gm, E, F♯m, A♯, and B. Activity in this same middle range for Hidden Unit 3 is produced by yet another subset of patterns: F, F♯m, F♯, Bm, and D♯. Finally, high activity (0.75 or greater) is produced in Hidden Unit 4 by F♯m, Dm, C♯m, Am, G♯m, F♯, Cm, and D.

What do all of these four different subsets have in common? One can take this question literally by examining the four different subsets of scales provided in the preceding paragraph: [C♯, Bm, F♯, B, F, A♯, F♯m], [Am, C♯m, Gm, E, F♯m, A♯, B], [F, F♯m, F♯, Bm, D♯], and [F♯m, Dm, C♯m, Am, G♯m, F♯, Cm, D]. To ask, "what do these sets have in common?" is to ask, "what is the intersection of the four subsets taken together?" By examining the four subsets, we discover that there is only one scale present in all four: F♯m. In other words, a stimulus that produces low activity in Hidden Unit 1, medium activity in Hidden Units 2 and 3, and high activity in Hidden Unit 4 must be a scale that has F♯ as its tonic and is of minor mode.

This account of how the four inaccurate hidden units can successfully isolate a single scale is an example of coarse coding. Coarse coding is an example of a distributed representation, which is one of the more interesting and important contributions of connectionism to cognitive science (Hinton, McClelland, & Rumelhart, 1986; Van Gelder, 1991). Coarse coding involves individual processors that are inaccurate detectors of some property or feature. For instance, a hidden unit might be inaccurate because it is broadly tuned, and can be activated either by a wide range of features or by wide range of levels of a specific feature (Churchland & Sejnowski, 1992; Hinton et al., 1986). Alone, such processors are not adept at solving classification problems.

However, if one combines the responses of many such poor detectors, then overall accuracy can be markedly enhanced. Typically, this requires that these poor detectors each have a slightly different view of the problem. That is, in coarse coding each processor is expected to evaluate different but overlapping ranges of stimuli. Each processor's response will be a combination of signal (the correct answer) and noise (additional incorrect responses). Processors with different perspectives on the problem will capture the same signal but will also likely capture different noise. When responses are pooled together, the different patterns of noise will cancel each other out, leaving only the correct answer. The intersection-of-subsets example above illustrates this general principle.

5.4.1 Implications

Our discussion of Figure 5-5 indicates that the hidden units transform the input pattern space via coarse coding. Discovering coarse coding in a network causes difficulties that were absent from the network interpretations we saw in previous chapters. A consequence of coarse coding is that it may be difficult, if not impossible, to relate network structure to formal music theory. It is very difficult to simply examine the patterns of bars in Figure 5-5 and easily discover musical structure. The reason for this is that coarse codes seem to be the antithesis of formal musical entities: they are fuzzy, messy, noisy, and poorly defined. Coarse coding is in fact one property of artificial neural networks that connectionists use to champion connectionist cognitive science as an alternative to classical cognitive science.

In the 1950s, classical cognitive science arose as a reaction against radical behaviourism (Bruner, 1990; Miller, 2003; Sperry, 1993). Behaviourism emphasized the study of the environment and of observable behaviour. In contrast, cognitivism argued that full accounts of psychological phenomena must appeal to mental representations and to the processes that manipulated them. The central claim of classical cognitive science is that cognition is computation (Pylyshyn, 1984).

Classical cognitive science has a particular view of computation: it is the kind of activity performed by a physical symbol system (Newell, 1980), a device like a modern digital computer. As a result, the mind is assumed to hold symbolic expressions that can be manipulated by the formal rules of a programming language called the language of thought (Fodor, 1975). Classical cognitive science is the current form of the logicist tradition (Boole, 1854/2003) that equated the laws of thought with the laws of formal logic. Given this view, it is far from surprising that a deeply formal system—music—has been extensively studied by cognitivists (Berkowitz, 2010; Deutsch, 1999; Howell et al., 1985; Krumhansl, 1990a; Sloboda, 1985; Temperley, 2001).

However, not all cognitive scientists agree that thinking is the rule-governed manipulation of mental representations (Dawson, 2013). While classical cognitive science believes that cognition is formal or symbolic, connectionist cognitive scientists disagree. For example, Paul Smolensky has argued that, in contrast to symbolic theories, artificial neural networks are subsymbolic (Smolensky, 1988; Smolensky & Legendre, 2006).

To say that a network is subsymbolic is to say that its individual hidden units do not detect interpretable features that could be represented as individual symbols. Instead, hidden units detect microfeatures. Individually, a microfeature is unintelligible, because its "interpretation" depends crucially upon its context (i.e., the set of other microfeatures that are simultaneously present [Clark, 1993]). However, a collection of microfeatures represented by a number of different hidden units can represent a concept that corresponds to a symbol in a classical model. In other words, the symbolic vocabulary of classical cognitive science is an approximate description of what emerges from finer-level mechanisms involving microfeatures, processor activities, and the like.

Clearly, coarse coding relates to Smolensky's notions of subsymbolic processes and microfeatures. However, if our analysis reveals coarse coding, then this likely means that we are discovering subsymbolic features that are much more difficult to relate to traditional music theory. This is a problem if our goal is to use network structure to provide insights about the theory of music, as we have been attempting in Chapters 3 and 4.

This is not to say that the coarse coding revealed in Figure 5-5 is not musical. The coarse coding of the hidden units supports musical classifications, and there are interesting musical properties revealed by the hidden unit space. For instance, an investigation of the distances between scales in the four-dimensional space (Figures 5-3 and 5-4) will reveal that different scales based upon the same tonic are near one another in this space. This is similar to the principle of placing a major key near its parallel minor in a Tonnetz (Schoenberg, 1969).

However, at times researchers might be tempted to emphasize the differences between subsymbolic musical networks and traditional music theory. That is, the extent to which the coarse codes of a network cannot be related to formal music may very well be the extent to which the network captures new regularities that cannot be expressed in formal theory. Importantly, the next chapter explores the possibility of an interesting compromise between hidden unit representations and formal music theory: we will explore networks that clearly exploit coarse coding, but these coarse codes are also related to formal music theory.

Before we move on to that material, though, let us consider a second approach to key-finding with neural networks. In the remainder of this chapter, we explore how one might create simpler neural networks that are variants of theories like the Krumhansl-Schmuckler key-finding algorithm (Krumhansl, 1990a). The purpose of these networks is to determine the keys of musical stimuli that are more complex than the simple scales that have been the focus, up to this point, of Chapter 5.

5.5 Key-Finding with Perceptrons

5.5.1 Key-Finding with Tonal Hierarchies

A critical component of listening to music is identifying its musical key. Human listeners are able to make such judgments very rapidly (Butler, 1989). Not surprisingly there is considerable interest in proposing procedures for musical key-finding, both to contribute to theories of music perception and to develop computer algorithms for automatically asserting the keys of musical stimuli (Albrecht & Shanahan, 2013; Frankland & Cohen, 1996; Handelman & Sigler, 2013; Holtzman, 1977; Longuet-Higgins & Steedman, 1971; Sapp, 2005; Shmulevich & Yli-Harja, 2000; Temperley, 1999; Temperley & Marvin, 2008; Vos & Van Geenen, 1996). An important feature of these theories is that they are intended for general musical stimuli, not just for the musical scales that were presented to the multilayer perceptron discussed in the preceding sections of Chapter 5.

In Section 5.1.1, we noted that the tonal hierarchy is one of the major findings in the study of musical cognition (Krumhansl, 1979, 1990a; Krumhansl & Shepard, 1979). Tonal hierarchies provide the foundation for one influential theory of key-finding proposed by Krumhansl and Schmuckler (Krumhansl, 1990a). The theory begins by recognizing that there is a different tonal hierarchy associated with each musical key. It uses these tonal hierarchies to create a set of key profiles, one for each of the 12 major and for each of the 12 minor musical keys. In the Krumhansl/Schmuckler key-finding algorithm, a to-be-analyzed musical stimulus is also represented as a key profile. This is accomplished by determining the total duration of each pitch-class in the stimulus. That is, one tabulates the total number of beats in the stimulus that involve hearing the pitch-class A, the total number of beats of the pitch-class A♯, and so on. Once the stimulus is represented in this fashion, correlations are computed between the stimulus's profile and each of the 24 standardized key profiles. The algorithm identifies the standardized profile that produces the highest correlation, and asserts that this is the key of the musical stimulus. Krumhansl (1990a) reports that this algorithm performs very well. It may also serve as a model of the cognitive processes involved when human listeners establish tonal centres (Frankland & Cohen, 1996; Schmuckler & Tomovski, 2005). For instance, Schmuckler and Tomovoski used the algorithm to predict listeners' experience of tonality for preludes by both Bach and Chopin.

Although influential and successful, the Krumhansl-Schmuckler algorithm is not problem-free. First, its performance is not perfect. For example, when examining the performance of the algorithm on a test set of 492 selections of classical compositions, Albrecht and Shanahan (2013) note that this procedure is only 74.2% accurate. Second, the performance of the algorithm varies depending upon whether it is presented stimuli in major or in minor keys. In Albrecht and Shanahan's test set, the Krumhansl-Schmuckler algorithm generated 69.0% accuracy for major key compositions, while it was 83.2% accurate in assigning minor keys.

Some researchers have investigated variations of the Krumhansl-Schmuckler algorithm in an attempt to improve key-finding performance. In general, these variations explore two different avenues. The first involves replacing the Krumhansl-Schmuckler tone profiles with alternative profiles derived from large corpuses of music pieces (Albrecht & Shanahan, 2013; Temperley, 2004). These alternative tone profiles are provided in Table 5-1 along with those used in the Krumhansl-Schmuckler algorithm.

Table 5-1 The three sets of key profiles used in key-finding algorithms.

	Scale degree of pitch-class											
	0	**1**	**2**	**3**	**4**	**5**	**6**	**7**	**8**	**9**	**10**	**11**
Krumhansl-Schmuckler												
Major	6.35	2.23	3.48	2.33	4.38	4.09	2.52	5.19	2.39	3.66	2.29	2.88
Minor	6.33	2.68	3.52	5.38	2.60	3.53	2.54	4.75	2.98	2.69	3.34	3.17
Temperley												
Major	0.75	0.06	0.49	0.08	0.67	0.46	0.10	0.72	0.10	0.37	0.06	0.40
Minor	0.71	0.08	0.48	0.62	0.05	0.46	0.11	0.75	0.40	0.07	0.13	0.33
Albrecht-Shanahan												
Major	0.24	0.01	0.11	0.01	0.14	0.09	0.02	0.21	0.01	0.08	0.01	0.08
Minor	0.22	0.01	0.10	0.12	0.02	0.10	0.01	0.21	0.06	0.02	0.06	0.05

Note. The major and minor profiles from Krumhansl (1990a), the major and minor profiles from Temperley (2004), and the major and minor profiles from Albrecht and Shanahan (2013). Scale degree 0 is assumed to be the tonic pitch, etc.

The second avenue for exploring variations of the Krumhansl-Schmuckler algorithm involves comparing inputs to standardized key profiles using some method other than correlation. Temperley (2004) uses the Bayesian probability equation, while

Albrecht and Shanahan (2013) assume that the standardized key profiles, and the to-be-classified stimulus, are all points located in 12-dimensional space; they then use the distance between pairs of points in this space to determine musical key.

5.5.2 A Key-Finding Perceptron

Let us now explore another variation of algorithms that use key profiles for key-finding. In this new approach, an artificial neural network learns to map different key profiles to outputs that represent musical key. Then we present the network a variety of different musical stimuli in order to test its ability to assert their musical keys.

Pilot studies reveal that a perceptron can accomplish this task (Figure 5-6). This perceptron uses 12 input units to represent key profiles, and uses 24 output units. Each of these output units represents a different musical key. Unlike the networks discussed up to this point, these output units are also integration devices that use the sigmoid-shaped logistic activation function instead of the Gaussian. I adopt these properties for two reasons. First, the output unit representation makes the behaviour of the network analogous to the behaviour of a correlation-based algorithm like the Krumhansl-Schmuckler. Second, using the logistic activation function permits us to interpret output unit activity as a probability such as the Bayesian probability employed by Temperley (1999).

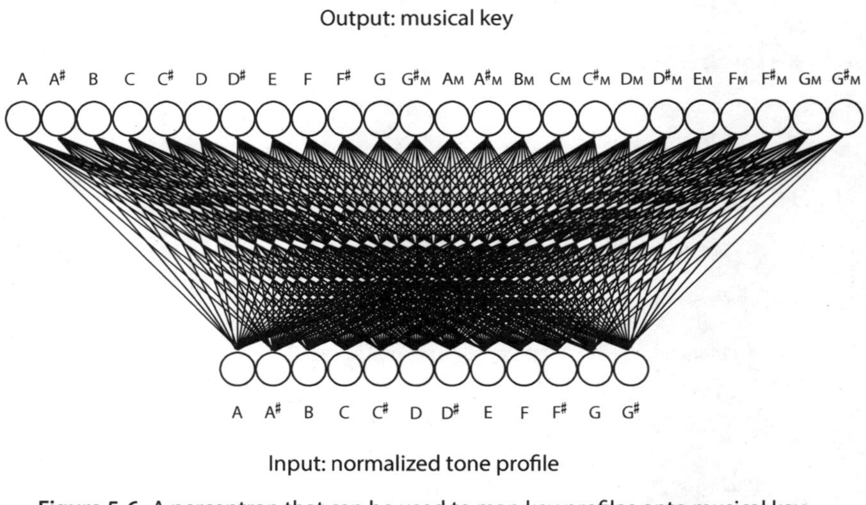

Figure 5-6 A perceptron that can be used to map key profiles onto musical key.

5.5.3 Perceptron Types

Three different types of key-finding perceptrons are trained. The first is trained on key profiles taken from the Krumhansl-Schmuckler algorithm (Krumhansl, 1990a), the second on key profiles taken from the Temperley (2004) algorithm, and the third on key profiles taken from the Albrecht and Shanahan (2013) algorithm.

Each of these three types of perceptrons represents variations of the original Krumhansl-Schmuckler correlational algorithm. Consider a perceptron trained on the Krumhansl-Schmuckler key profiles. It differs from the original Krumhansl-Schmuckler algorithm in two ways. First, it does not directly match key profiles to musical keys. Instead, it uses its connection weights to transform key profiles before their signals reach the output units. Second, this network does not use correlations to match stimuli with desired musical keys. Instead, the output units take the signals from (transformed) input profiles and then further transform them by applying the logistic equation. It is therefore possible that a perceptron trained on the Krumhansl-Schmuckler key profiles will respond differently to the same stimuli in comparison to the Krumhansl- Schmuckler algorithm itself.

5.5.4 Training Sets

One advantage that the Krumhansl-Schmuckler algorithm has via its use of correlations to compare stimuli to key profiles is that this operation is not affected by stimulus magnitude. That is, when a long musical selection is summarized as a pitch-class profile, the magnitude of each value in its 12-dimensional vector is expected to be larger than would be observed when a shorter musical selection is summarized in the same way. This is simply because on average one would expect to find more instances of each pitch-class in a longer piece than in a shorter one. However, the correlation equation is not sensitive to stimulus magnitude. Instead, it in essence computes the similarity of two patterns by only considering their relative directions (when they are considered as vectors pointing in a multidimensional space).

As perceptrons do not use correlations, their outputs can be affected by differences in stimulus magnitude. For this reason, I first mean-centre and then normalize each key profile from Table 5-1. The resulting normalized key profiles associated with each of these three different algorithms are provided in Table 5-2.

Table 5-2 The three sets of mean-centred normalized key profiles used to train different key-finding perceptrons. These are the same profiles that were presented earlier in Table 5-1, but each has been processed as described in the text.

	Scale degree of pitch-class											
	0	1	2	3	4	5	6	7	8	9	10	11
Krumhansl-Schmuckler												
Major	0.655	−0.286	−0.001	−0.263	0.205	0.139	−0.220	0.390	−0.250	0.041	−0.273	−0.138
Minor	0.668	−0.234	−0.026	0.433	−0.253	−0.024	−0.268	0.278	−0.160	−0.231	−0.071	−0.113
Temperley												
Major	0.442	−0.330	0.151	−0.305	0.355	0.119	−0.289	0.405	−0.280	0.014	−0.333	0.052
Minor	0.423	−0.308	0.146	0.313	−0.348	0.130	−0.283	0.463	0.064	−0.327	−0.251	−0.022
Albrecht-Shanahan												
Major	0.575	−0.287	0.103	−0.287	0.199	0.040	−0.250	0.485	−0.276	−0.012	−0.280	−0.009
Minor	0.563	−0.318	0.086	0.164	−0.264	0.082	−0.293	0.538	−0.087	−0.252	−0.091	−0.128

To build a training set for a perceptron trained on the Krumhansl-Schmuckler profiles, the mean-centred normalized major key profile in the first row of Table 5-2 is used to generate a key profile for each of the 12 major keys by associating the value given in Table 5-2 with the appropriate input unit. That is, for the key of C major the value of 0.655 is presented to the input unit representing the pitch-class C, the value of −0.286 is presented to the input unit representing the pitch-class C♯, and so on. Similarly, the stimulus for the key of C♯ major involves presenting the value of 0.655 to the input unit representing the pitch-class C♯, the value of −0.286 to the input unit representing the pitch-class D, and so on. A similar procedure creates input stimuli for the 12 different minor keys using the normalized minor key profile in the second row of Table 5-2. As a result, the training set consists of 24 different input patterns, one for each musical key. A second set of 24 training patterns is created by applying this method with the two Temperley profiles from Table 5-2, and a third set of 24 training patterns is created by applying this method using the two Albrecht-Shanahan profiles from Table 5-2.

5.5.5 *Training*

Each network is a perceptron of the type illustrated in Figure 5-6, and is trained using the Rosenblatt software program (Dawson, 2005). Before training begins, all connection weights in a network are set to random values between −0.1 and 0.1. I initialize all the biases (θ) of the output units to zero, but I then modify them using the learning rule during training. I employ a learning rate of 0.50. Networks are trained epoch by epoch, where each of the 24 training patterns is presented once each epoch. The order of pattern presentation is randomized each epoch. Training continues until the network generates a "hit" for every output unit for each of the 24 patterns in the training set. A hit is defined as an activity of 0.90 or higher when the desired output is one or as an activity of 0.10 or lower when the desired output is zero.

Each network is trained from a random configuration of a small initial set of connection weights, it is possible that different networks will achieve qualitatively different end states via training. For this reason, I train 10 different perceptrons on each of the three sets of training patterns, resulting in 30 different perceptrons. Each of these perceptrons is a different "subject" in a simulation experiment.

All 30 perceptrons successfully learned to map mean-centred normalized tone profiles onto musical keys. On average the 10 perceptrons trained on the Krumhansl-Schmuckler profiles converge after 1354.3 epochs of training (*SD* = 0.823). On average the 10 *perceptrons* trained on the Temperley profiles converge after 1453.4 epochs of training (*SD* = 1.17). On average the 10 perceptrons trained on the Albrecht-Shanahan profiles converge after 1864.0 epochs of training (*SD* = 1.25).

As 10 different versions of each perceptron were trained, and as each of these began from a different set of small randomly selected initial weights, are there any qualitative differences between the solutions reached by different versions of the same perceptron type? Interestingly, it appears that each perceptron trained on the same set of profiles reaches essentially the same solution (i.e., the same set of connection weights, as is detailed later) and generates essentially the same responses to stimuli.

5.5.6 *Testing*

After successfully completing the training phase, a network is tested on its ability to assert the musical key of 296 different musical selections. These test stimuli represent four different sources. The first is the collection of 48 preludes and 48 fugues from both books of J.S. Bach's Well-Tempered Clavier. These compositions are a typical test bed for key-finding algorithms (Albrecht & Shanahan, 2013; Temperley,

1999). The second is a collection of 24 preludes composed by Frederic Chopin as his Opus 28. One prelude was written in each musical key. These too are often used to test the accuracy of key-finding theories; for instance, they pose some difficulty for the Krumhansl-Schmuckler algorithm (Krumhansl, 1990a). The third test set is the 24 Preludes, Opus 67, composed by Johann Hummel. As is the case for the Chopin preludes, Hummel composed one prelude in each musical key. The fourth test set, the only selection of music not from the classical genre, consists of 152 Nova Scotia folk songs from *Songs and Ballads from Nova Scotia* (Creighton, 1932).

Each of these four collections of musical selections is available in the kern file format at http://kern.humdrum.org/. As a result, they can be analyzed using the Humdrum toolkit (Huron, 1999). Humdrum's key command is used to represent each of the 296 test stimuli in the format required by the Krumhansl-Schmuckler algorithm. That is, the total duration of each of the 12 Western pitch-classes is computed for each stimulus, producing a 12-entry vector representation. Each of these vectors is then normalized to unit length prior to being presented to a trained network. This normalization renders the test stimuli into a format that is identical to the one used to represent the training stimuli. It also ensures that the varying lengths of each of these test patterns do not affect network performance.

During the test phase, network performance is assessed using a procedure analogous to the Krumhansl-Schmuckler algorithm. That is, when a test stimulus is presented to a network it produces activity in 24 different output units, each one representing a different musical key. The output unit that generates the highest activity is taken to indicate the musical key being asserted by the network. The accuracy of this assertion is then compared to the key in which the stimulus was actually composed (information that is provided as part of the test stimulus's kern file).

5.5.7 Perceptron Performance

The first row of Table 5-3 provides the average accuracy of key assertion for perceptrons trained on the Krumhansl-Schmuckler profiles for the four different sets of test materials. For each set, accuracy is given as the percent correct for the entire set of stimuli; accuracy is then provided for only the major key stimuli and for only the minor key stimuli. The next two rows of Table 5-3 then provide the performance for perceptrons trained with the Temperley profiles and for perceptrons trained with the Albrecht-Shanahan profiles.

In order to have some reference point for assessing perceptron performance, the Krumhansl-Schmuckler algorithm is also employed. The fourth row of Table 5-3 provides key-finding accuracy when correlations between test stimuli and the normalized Krumhansl-Schmuckler profiles are used (where key is asserted by finding the highest correlation).

Table 5-3 The average percent accuracy of classification of the three perceptrons trained on three different mean-centred and normalized key profiles.

Key-finding method	Bach Well-Tempered Clavier			Chopin preludes			Hummel preludes			Nova Scotia folk songs		
	All	Major	Minor	All	Major	Minor	All	Major	Minor	All	Major	Minor
KSP	89.6	89.6	89.6	54.2	75.0	33.3	87.5	100.0	75.0	66.5	70.4	35.3
TP	95.8	97.9	93.8	95.8	91.7	100.0	91.7	91.7	91.7	70.4	78.5	5.9
ASP	96.9	93.8	100.0	83.3	66.7	100.0	100.0	100.0	100.0	74.3	79.3	35.3
KS Corr	93.8	87.5	100.0	87.5	83.3	91.7	100.0	100.0	100.0	67.1	71.9	29.4

Note. KSP indicates the perceptron trained on the Krumhansl-Schmuckler profiles; TP indicates the perceptron trained on the Temperley profiles, and ASP indicates the perceptron trained on the Albrecht-Shanahan profiles. The final row (KS Corr) provides the performance of the Krumhansl-Schmuckler correlation algorithm for purposes of comparison.

One observation to make from Table 5-3 is that perceptrons trained on tone profiles demonstrate different key-finding performance for stimuli in major or minor keys. In some instances, a perceptron is better on minors than on majors, while in other instances the reverse is true. Interestingly, the perceptron that performs best on minor key stimuli is the one trained on the Albrecht-Shanahan profiles; for classical genre patterns, it is 100% accurate. One of the motivations that Albrecht and Shanahan (2013) provided for their profiles was the goal of improving key-finding for minor key stimuli.

Another observation to make from Table 5-3 is that performance on classical genre stimuli is much better—for both perceptrons and for the correlation algorithm—than it is for the Nova Scotia folk songs. This may be due to a variety of factors. For instance, the folk songs are generally short univocal selections, while the classical pieces are generally longer and include harmony. As a result, there may be more reliable information about key in the classical selections than in the folk songs. Table 5-3 indicates that particular sets of tone profiles for key-finding might have more success for some genres, or for at least some subsets of stimuli, than for others.

One final observation to make concerning Table 5-3 is that when the same set of profiles is used, but processed differently, the same performance is not produced. In particular, the Krumhansl-Schmuckler perceptron generates significantly different levels of accuracy than the Krumhansl-Schmuckler correlation algorithm, even though both of these methods use the same profiles. This shows that the perceptron is processing the profiles in a different fashion than when correlation is used.

The results of this study indicate that a very simple artificial neural network, the perceptron, is capable of mapping key profiles onto musical keys, and can then use this learned ability to perform respectably when given the task of asserting the keys of novel stimuli. In addition, it is clear that the outputs of a perceptron trained on key profiles are not identical to the judgments of an algorithm that asserts musical key using correlations with the same key profiles. This difference provides further support for the position that key-finding performance can be altered by changing the method used to compare the key profile of a stimulus with the key profiles associated with different musical keys. Finally, perceptron performance is clearly affected by which key profiles are used for training. Being trained on the normalized Temperley profiles leads to better performance when presented novel stimuli than being trained on the normalized Krumhansl-Schmuckler profiles.

5.6 Network Interpretation

5.6.1 Interpreting Perceptron Weights

The key-finding perceptrons that I have been discussing differ from the previous networks in the use of the logistic activation function in their output units. Earlier I noted that, in general, networks that use value units are easier to interpret than networks that use integration devices. However, when the network has no hidden units, a quantitative analysis of an integration device network's structure can be carried out very easily. This is because a perceptron that uses integration devices as outputs is functionally equivalent to a system that performs logistic regression (Schumacher, Rossner, & Vach, 1996). This, in turn, means that a network's weights can be interpreted as being natural logarithms of odds ratios, and that the size of each weight provides a measure of the importance of each input with respect to activating an output unit (Dawson & Gupta, 2017). When a perceptron uses integration devices as outputs, its connection weights literally reflect the effect size of each input signal—the degree to which each part of a stimulus is responsible for turning an input unit on or off.

In Chapter 3, instead of viewing input units in terms of the pitch-class they represented, I considered them in terms of their degree within a scale related to an output unit (coding the input unit representing the scale's tonic pitch-class as 0, coding the input unit representing the pitch-class a semitone higher than the tonic as 1, and so on). This is how I transformed Table 3-2 into Table 3-3 when interpreting the structure of the scale tonic perceptron. I found that, with this recoding, each output unit had essentially the same set of connection weights feeding into it.

This is also the case when this recoding is performed for each of the three key-finding perceptrons. That is, for each perceptron, each of its output units that represents a major key has the same set of input weights feeding into it; a different pattern of weights is found for each of the network's output units that represents a minor key. Table 5-4 provides the connection weights that I discovered for each perceptron, and for each key type, when input units were considered in terms of a pitch's relation to the key (with respect to degree) instead of absolute pitch-class.

Table 5-4 Weights from input units to typical output units of the three perceptrons.

	Weight between input unit (coded in terms of key degree) and output unit												
	0	**1**	**2**	**3**	**4**	**5**	**6**	**7**	**8**	**9**	**10**	**11**	**θ**
Krumhansl-Schmuckler													
Major	6.68	−0.63	−1.63	−3.78	3.57	1.57	0.39	3.44	−1.60	−2.07	−3.45	−2.39	−7.15
Minor	6.46	0.11	0.41	4.61	−3.35	−1.69	−0.67	0.21	−2.84	−2.30	−2.22	1.28	−6.05
Temperley													
Major	3.67	−0.76	1.41	−2.78	5.67	1.77	−2.57	4.52	−2.57	−3.63	−3.92	−0.77	−7.47
Minor	3.13	−1.71	1.80	3.30	−4.14	1.65	−0.50	2.76	0.60	−3.03	−5.57	1.74	−6.14
Albrecht-Shanahan													
Major	7.27	0.48	−0.95	−5.04	4.84	−0.78	−1.13	4.33	−2.46	−2.44	−3.77	−0.32	−8.28
Minor	5.73	−2.20	0.46	4.08	−5.66	−0.32	−0.59	5.05	−0.70	−4.06	−2.97	1.13	−7.94

Note. Input units are coded in terms of key degree; biases are provided in the column labelled θ. ASP indicates the perceptron trained on the Albrecht-Shanahan profiles, TP indicates the perceptron trained on the Temperley profiles, and KSP indicates the perceptron trained on the Krumhansl-Schmuckler profiles. Unit type indicates whether the output unit represents a major key or a minor key.

In all of the key-finding algorithms that use key profiles (Albrecht & Shanahan, 2013; Krumhansl, 1990a; Temperley, 2004), there is a direct comparison of the profile representing a musical stimulus to the profile of each musical key. That is, the tacit assumption underlying these algorithms is that the key profile itself (Table 5-1) provides enough information to identify key. If the different values of a key profile were themselves sufficient to select the correct key, then a perceptron would not have to weight these input values when learning to map these profiles into musical keys. That is, each of its connection weights would equal 1, because the perceptron would not have to alter its stimulus information. However, Table 5-4 reveals that this is clearly not the case: Table 5-4 contains connection weights ranging from over 7 to about –6. This indicates that different components of stimulus profile have different degrees of importance in determining musical key.

It is important to realize that these Table 5-4 weightings of input unit signals are in addition to the different-sized signals that are reflected in the profiles themselves. For instance, the mean-centred normalized Albrecht-Shanahan major key profile itself indicates that the most common component in the profile is degree 0, because its profile value is 0.575. However, the perceptron further amplifies this value by multiplying it by 7.29, indicating that this high profile value is itself extremely informative relative to the other profile values. Similarly, in the Albrecht-Shanahan major profile the values of degree 1 and degree 3 both equal –0.29. However, the perceptron weights these two values quite differently, multiplying the degree 1 signal by a weight of 0.44, while multiplying the degree 3 signal by a weight of –5.04. This indicates that the perceptron has learned that decreased occurrence of degree 3 pitch-classes is far more important for establishing the major key than is decreased occurrence of degree 1 pitch-classes. Similar observations can be made for the various signals contributing to the identification of either major or minor keys, and these observations can be made for all three types of perceptron.

5.6.2 Implications

Table 5-4 indicates that when key profiles are used to train a key-finding perceptron, it learns that not all components of the profile are equally important. Perceptrons adjust their weights to modify the input signals to emphasize the information provided by important profile elements, and to de-emphasize the information provided by less important elements. If this were not the case, then all of the weights in Table 5-4 would equal 1. One interesting implication of this finding is that the structure of a trained perceptron suggests possible variations of other key-finding algorithms that do not employ artificial neural networks.

For example, the Albrecht-Shanahan algorithm positions each tone profile as a point in a 12-dimensional space, positions a stimulus profile in the same space, and assigns the stimulus the key that is represented by the point the stimulus is closest to (Albrecht & Shanahan, 2013). However, this model assumes that each dimension in the space has the same importance. The Albrecht-Shanahan perceptron weights (Table 5-4) indicate that the space in which the Albrecht and Shanahan algorithm measures distances could be distorted, with some dimensions being stretched out (i.e., those associated with important profile components), and with others being shortened in size (i.e., those associated with less important profile components). The weights of the perceptron provide magnitudes for distorting the Albrecht-Shanahan space before points are plotted and distances are measured; it would be interesting to see what effect such distortions would have on the algorithm's performance. A similar approach could be taken to incorporate weights into other algorithms that are based on tone profiles by using them to scale profile components. Of course, the weights in Table 5-4 would likely not be the precise ones to use, because they arise from networks trained on mean-centred normalized tone profiles. However, a perceptron trained to map musical keys to tone profiles that have not been preprocessed (i.e., those presented earlier in Table 5-2) could provide weights that could be incorporated into another algorithm.

5.7 Summary and Implications

5.7.1 Summary

This chapter examined key-finding from two different perspectives. I began by exploring a multilayer perceptron for identifying the tonic and mode of scales. I then discovered that it developed a different kind of representation, a coarse code, for determining scale tonics and modes.

For the second perspective, we trained a perceptron to key-find using key profiles taken from established theories of key-finding. After this training, I tested the perceptron using a variety of different compositions after summarizing them in the way required by the original Krumhansl-Schmuckler algorithm. In general, perceptrons trained on key profiles are plausible algorithms for asserting musical key. Furthermore, an examination of the weights of these trained networks indicates that some components of a key profile are more important than others with respect to performing key-finding. Let us now consider some of the implications of the results observed in this chapter.

5.7.2 *Coarse Coding*

One of the major themes of this book is that artificial neural networks provide a medium for discovering new mappings between inputs and outputs. This theme was illustrated in our interpretations of a tonic-identifying perceptron in Chapter 3, and in our interpretations of a mode-identifying multilayer perceptron in Chapter 4. The first half of Chapter 5 demonstrated that when a multilayer perceptron learns to accomplish both of these tasks at the same time it discovers a new method that utilizes coarse coding.

The discovery of coarse coding has some interesting implications for the study of musical cognition. The general perspective of this field of study is that mental representations are used to organize and understand musical stimuli. However, the nature of these mental representations is an open question. One possibility is that geometric or spatial representations used to model musical regularities might also serve as accounts of mental musical representations (Krumhansl, 1990a, 2005).

Coarse coding provides an alternative representational possibility. Networks that employ coarse coding should exhibit particular patterns of learning (i.e., learning some instances faster than they learn others), should make particular patterns of responses to novel stimuli, and should generate definite errors when presented noisy stimuli. One might ask whether human listeners use coarse codes to represent music, and search for an answer to this question by comparing patterns of human responses under various conditions to related patterns observed in networks that are known to coarse code.

Coarse coding has less obvious, but still definite, implications for formal music theory. Our interpretation of Figure 5-5 focused on a formal process (intersections of subsets) that could use the coarse code to identify a scale's tonic and mode. This formal process is carried out when an output unit's activation function carves decision regions in a hidden unit space.

Our interpretation of each hidden unit's response (Figure 5-5) was cursory, and did not explore particular musical features. I simply noted that the patterns of activity depicted in that figure summarize "subsymbolic" properties. However, such properties, though subsymbolic, are almost certainly going to be musical in nature. A more intensive (and likely more challenging) examination of hidden unit responses could identify the musical properties that cause a hidden unit to generate high responses to some stimuli but not to others. These properties would likely represent an alien music theory. They would also be important for informing representational theories of music cognition.

5.7.3 *Asserting Musical Key*

One of the important results of the second half of Chapter 5 is that perceptrons trained on key profiles from other theories can perform quite well when identifying the keys of novel stimuli. These results are very encouraging, because they indicate that artificial neural networks demonstrate a great deal of promise as key-finding mechanisms.

These results are also surprising. First, the fact that simple perceptrons are capable of this high degree of performance is unexpected. Second, this high degree of performance is produced when a very sparse representation of input stimuli is employed. For instance, given the various modulations of key that characterize the pieces that make up the Well-Tempered Clavier (Bruhn, 2014; Temperley, 1999), it is surprising that a 12-number summary of each piece permits a perceptron to judge their intended key (as per Bach's titling) at nearly 98% accuracy. If anything, the success of these simple networks provides converging validation about the fundamental importance of Krumhansl's (1990a) tonal hierarchy.

The three different perceptrons also point to even further avenues of exploration. First, the three sets of tone profiles (Albrecht & Shanahan, 2013; Krumhansl, 1990a; Temperley, 1999) used to train the perceptrons are derived from different origins, are typically employed using different algorithms, and involve different magnitudes of values (as can be seen by inspecting Table 5-1). Given these differences, it is perhaps not surprising that no one has attempted to improve key-finding by combining all three sets of profiles together in a single algorithm.

However, from the perspective of artificial neural networks, combining different profiles together is a natural next step. Indeed this notion follows naturally from the notion of coarse coding, if we consider coarse coding not in terms of hidden units but instead in terms of perceptron responses. Table 5-3 indicates that the responses of different types of perceptrons differ from one another on the various test stimuli. This suggests that each captures slightly different properties. From the coarse coding perspective of neural networks, this in turn suggests that key-finding performance could be improved in an algorithm that combines the three different sets of tone profiles together.

There is no reason that a single perceptron cannot be trained to key-find by learning about all three different types of key profiles in a single training set. Second, and more in the spirit of coarse coding, one could assert musical key after combining the responses of different perceptrons together. Such an architecture is called a committee of networks; committees of networks have been shown to be superior to single networks in a variety of pattern classification tasks (Buus et al., 2003; Das,

Reddy, & Narayanan, 2001; Guo & Luh, 2004; Marwala, 2000; Medler & Dawson, 1994; Zhao, Huang, & Sun, 2004). It would be interesting to see how a committee of networks performs on the key-finding task in comparison to the perceptrons that have been described in this chapter.

6

Classifying Chords with Strange Circles

6.1 Four Types of Triads

Up to this point, most of the musical networks considered have explored tonality in terms of components related to melody, such as the pitch-classes that make up a particular scale. I will now explore networks that are faced with making judgments involving harmony, where multiple stimulus elements combine together to define a musical combination like a chord. I begin by defining basic chords called triads, and interpret the structure of a network that learns to classify triads by their quality. Next, I discover a property called strange circles, in which hidden units in a network organize pitch-classes into musical systems called interval cycles, but then treat each member of an interval cycle as being the same pitch. Then, I turn to defining, geometrically, all of the possible interval cycles in Western music in order to have a catalog of strange circles available for network interpretation. Finally, I define another kind of chord, the added note tetrachord, and train a network to classify stimuli into different types of such chords. When this network is interpreted, I once again discover that the network uses strange circles.

6.1.1 Triads

The simulations described in Chapter 3, Chapter 4, and the first part of Chapter 5 involved training artificial neural networks to identify properties of a particular type of monophonic music, the scale. In this chapter, I turn to training networks to detect some basic properties of harmonic music, where combinations of inputs are of key import. To begin, I focus on a simple chord, the triad, which is composed of three distinct notes. The most basic form of triad is the major triad, created by combining the first, third, and fifth notes of a major scale. Consider the A major scale that was presented in Chapter 3 as Figure 3-1. Its first note is A, its third note is C♯, and its fifth note is E. Therefore the A major triad, symbolized as A, is the notes A-C♯-E. Figure 6-1 illustrates this triad in the very first bar of the score.

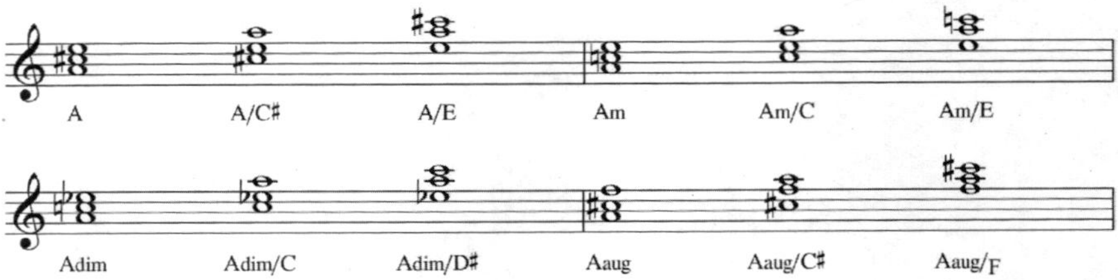

Figure 6-1 The top staff provides examples of four different types of triads built on the root note of A: A major (A), A minor (Am), A diminished (Adim) and A augmented (Aaug).

When the triad's notes are placed on a musical staff such that the first note of the major scale is the lowest note, the third note is the next lowest, and the fifth note is the highest, the triad is said to be in root position. The A triad at the top left of Figure 6-1 is in root position. The major triad can be manipulated to create other types of triads that have different sonority. For instance, if one takes the middle note of a major triad and lowers it by a semitone, the result is a minor triad. The second bar of the Figure 6-1 score provides the A minor (Am) triad in root position; it is composed of the notes A-C-E. Note that an alternative way to construct a minor triad is to combine the first, third, and fifth notes of a minor scale. For instance, A, C, and E are the first, third, and fifth notes of the A harmonic minor scale presented earlier in Figure 3-1.

A different alteration of the major triad produces a third kind of triad, the diminished triad. A diminished triad is produced by lowering both the second and the third notes of a major triad by a semitone. The A diminished triad (Adim) is shown in the third bar of the Figure 6-1 score, and consists of the notes A-C-E♭. A third alteration of the major triad produces yet another triad, the augmented triad. An augmented triad is created by taking a major triad and raising its third note by a semitone. The augmented triad for A (Aaug) is illustrated in root position in the fourth bar of the Figure 6-1 score. The first four bars of Figure 6-1 provide four different triads built upon the root of A. Obviously major, minor, diminished, and augmented triads can be built using any of the 12 possible Western music pitch-classes as the root. Therefore, there are 12 different versions of each of these four different types of triad. Later in this chapter we will explore a multilayer perceptron that has the task of identifying triad type, ignoring the root or key of the triad.

6.1.2 Inversions and Representations

Triads like those presented in the first line of Figure 6-1 can take different forms through a process called inversion. To create the first inversion of a triad in root position, one raises its lowest note an octave. To create the second inversion of a triad in root position, one raises both its lowest and its middle notes an octave. The first and second inversions of each of the triads are shown in Figure 6-1.

Inverting a triad does not alter the triad's quality. As a result one interesting task is to train a network to identify triad quality (major, minor diminished, augmented) regardless of the triad's key and regardless of the triad's inversion. However, to accomplish this task I must adopt a different input representation than the one used to represent scales in Chapters 3 through 5. All of these networks used 12 input units to represent stimuli in terms of their component pitch-classes. However, in order to present different inversions of the same triad I must replace one pitch with another pitch that is an octave higher. This distinction between pitches an octave apart is not possible with the pitch-class representation that has been used for the earlier networks, because it representation assumes octave equivalence.

In order to present different inversions of the same chord, I must replace the pitch-class representation with a local representation of musical pitch. In this local representation, each input unit can be considered analogous to a key on a piano that is associated with a particular pitch. For instance, one input unit might represent the pitch A3 (the A below middle C) while a different input unit represents the pitch A4 (the A above middle C). Both of these units represent the same pitch-class (A), but different pitches.

Figure 6-2 illustrates a multilayer network for triad classification that uses a local representation of pitch in order to present triads in different inversions. Its 28 different input units represent 28 different musical pitches. The input units of this network are analogous to a 28-key piano keyboard whose lowest key plays the note A3 (the A below middle C), and whose highest key plays the note C6 (the C two octaves higher than middle C).

6.2 Triad Classification Networks

6.2.1 Task

The task for the multilayer perceptron described in this section is to identify triad type (major, minor, diminished, or augmented), ignoring both the triad's key and its inversion.

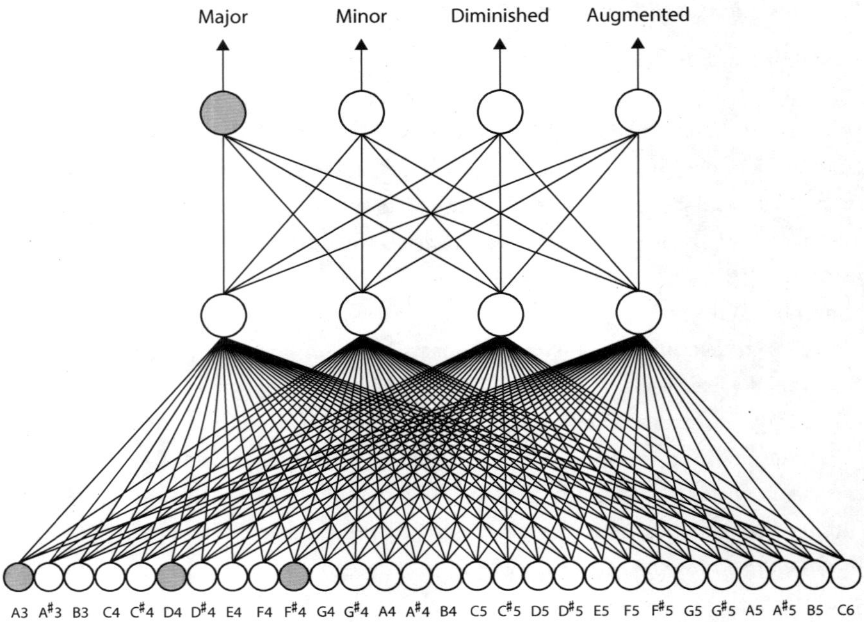

Figure 6-2 A multilayer perceptron with local pitch encoding that learns to identify four types of triads, ignoring a triad's key and inversion.

6.2.2 Network Architecture

The multilayer perceptron illustrated in Figure 6-2 employs four output units. Each is a value unit. This network requires four hidden value units to converge to a solution to the triad classification problem. The network uses local encoding of specific pitches in order to present different inversions of triads as was discussed in Section 6.1.2. This input unit layout provides enough "room" to encode the three different forms of each triad. The lowest triad presented to the network is A major in root position. The highest triad presented to the network is G♯ augmented in second inversion.

6.2.3 Training Set

The training set consists of 144 stimuli: 36 different major triads, 36 different minor triads, 36 different diminished triads, and 36 different augmented triads. The four different triad types are constructed using the 12 different root pitch-classes that were available. In addition, each type of triad, for each root, is created in three different forms (see Figure 6-1): root position, first inversion, and second inversion. A particular triad (i.e., a triad related to a particular root, in a particular form) only appears once in the training set. For instance, there is only one A major triad in root position (whose lowest note is the first of the 28 input units), even though this same triad could be presented using other input units higher up along the network's "keyboard."

Each triad is encoded as an input pattern in which three input units are activated with a value of one, and the remaining 25 input units are all activated with a value of zero. Each input pattern is paired with an output pattern that requires one output unit to activate with a value of one, and the other three output units to activate with a value of zero. The output unit trained to activate is one that represents the input pattern's correct triad type.

6.2.4 Training

The multilayer perceptron is trained using the generalized delta rule developed for networks of value units (Dawson & Schopflocher, 1992) that is part of the Rumelhart software program (Dawson, 2005). All connection weights in the network are set to random values between −0.1 and 0.1 before training begins. The μs of the output and hidden units are all set to zero throughout training. We employ a learning rate of 0.005. Training proceeds until the network generates a "hit" for every output unit for each of the 144 patterns in the training set. Once again, a "hit" is defined as activity of 0.9 or higher when the desired response is one or as activity of 0.1 or lower when the desired response is zero. The multilayer perceptron typically learns to solve this problem in fewer than 1000 epochs of training. The example network described in more detail below converged after only 576 sweeps of training.

6.2.5 Connection Weight Patterns

This triad classification task produces some very interesting properties when the connection weights of a trained network are examined. To begin, let us examine the connection weights from each of the 28 input units to Hidden Unit 1 of this multilayer perceptron, which appear in Figure 6-3. A quick glance at this figure reveals a striking regularity in connection weight patterns. Ignoring slight variance in connection weight magnitude, there is a pattern that repeats itself every three input units: a weak negative connection weight, followed by a strong negative connection weight, followed by a strong positive connection weight.

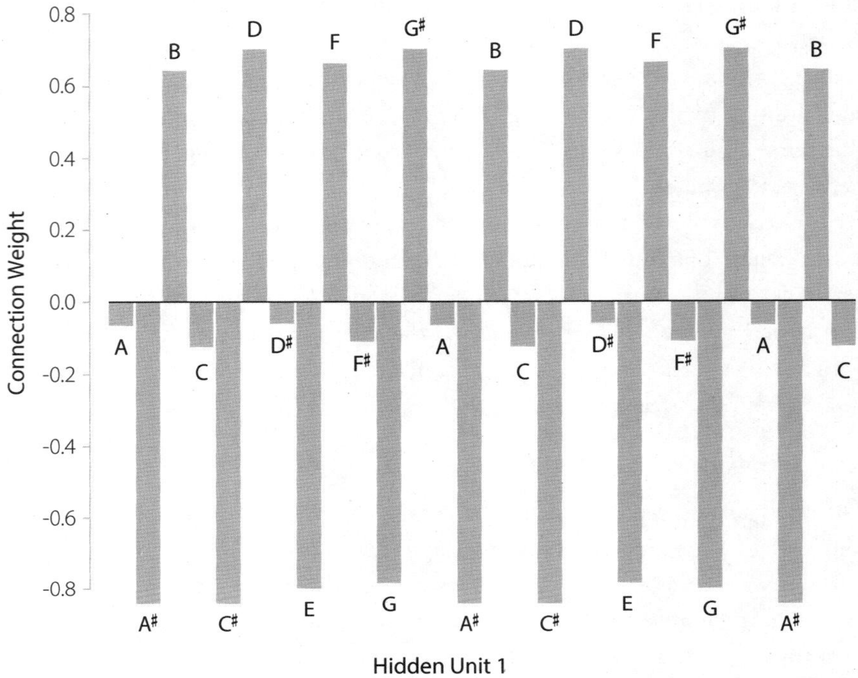

Figure 6-3 The connection weights from the 28 input units for pitch to Hidden Unit 1 in the network trained to classify triad types for different keys and inversions.

This repeating pattern illustrates a number of different equivalences between pitch-classes separated by a specific musical interval. The first is octave equivalence: every input unit that represents a pitch that belongs to the same pitch-class (e.g., the three different A pitches) is assigned the same weight. Second, pitch-classes separated by the interval of a minor third (three semitones) are all assigned roughly the same weight. For instance, B, D, F, and G♯ all belong to the interval cycle of minor thirds; that is, each of these pitches is a minor third away from its adjacent neighbours in the list. (The properties of such interval cycles [Roig-Francolí, 2008] are detailed later in this chapter.) Critically, they all have approximately the same connection weight in Figure 6-3. Similarly A, C, D♯, and F♯ belong to a different interval cycle of minor thirds; they too are assigned the same connection weight (weak negative), but one that is different from the weight assigned the first set of pitches. Finally A♯, C♯, E, and G all belong to yet a third interval cycle of minor thirds; they are also assigned the same connection weight (strong negative), but one that distinguishes them from the two other groups of pitches.

Paying more attention to the subtle differences in connection weights in Figure 6-3, one also finds evidence for tritone equivalence. For instance, A and D♯ have nearly identical connection weights, as do C and F♯. All pitches related by tritones that belong to the same "circle of tritones" are assigned the same weight. In other words, this set of connection weights indicates that pitches that are specific musical intervals apart from one another are assigned nearly identical connection weights. For Hidden Unit 1, this means that these different pitches are in reality all the same, because the connection weight from an input unit to the hidden unit in essence can be interpreted as Hidden Unit 1's "name" for the input pitch.

Figure 6-4 presents the connection weights from the input units of this network to a different hidden unit, Hidden Unit 2. It too reveals some striking interval equivalences, some of which differ from those seen in Figure 6-3. In Figure 6-4, a general pattern of connection weights repeats itself every four input units instead of every three. This pattern is a weak negative weight, then a strong positive weight, then a weak positive weight, and finally a strong negative weight.

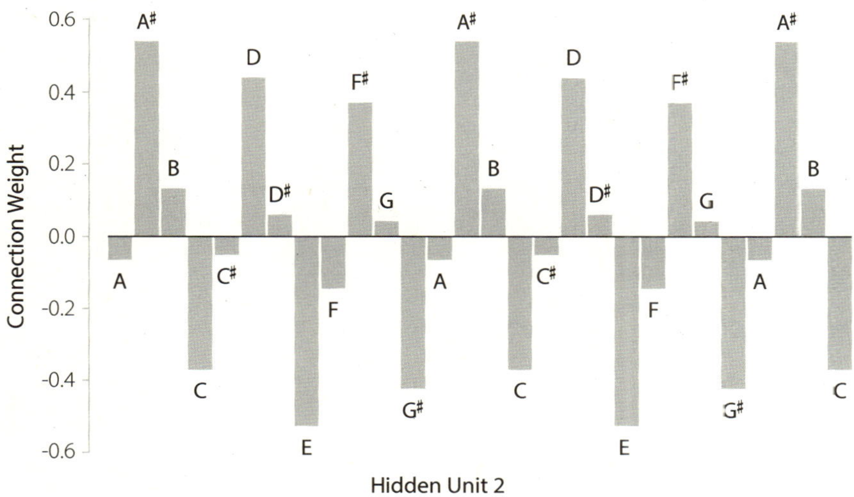

Figure 6-4 The connection weights from the 28 input units for pitch to Hidden Unit 2 in the network trained to classify triad types for different keys and inversions.

Once again, this pattern exhibits octave equivalence: each pitch that belongs to the same pitch-class has the same weight. Ignoring subtle variations in magnitude, the pattern of weights in Figure 6-4 separates the different pitches into four different groups, each containing three pitch-classes. For example A, C♯, and F are the only pitches that have weak negative weights. Analogous groupings via connection weights are found for the pitch-classes A♯, D, and F♯, for the pitch-classes B, D♯, and G, and for the pitch-classes C, E, and G♯.

The weights in Figure 6-4 do not exhibit tritone equivalence but do demonstrate tritone balance: note that pitch-classes separated by a tritone (like A and D♯) have weights that are equal in magnitude but opposite in sign.

Figure 6-5 presents the connection weights between the input units and Hidden Unit 3. Its organization is very similar to that of Hidden Unit 2 (Figure 6-5). It exhibits octave equivalence, organizes pitches into the four different groups related by major thirds, and demonstrates tritone balance. The key difference between Figure 6-5 and Figure 6-5 is that while both exhibit the same organizational pattern, there appears to be a phase shift of this pattern when we compare the two hidden units. That is, the weak negative weight that starts the pattern is associated with A in Figure 6-4 but has been shifted three semitones to the right to start with C in Figure 6-5.

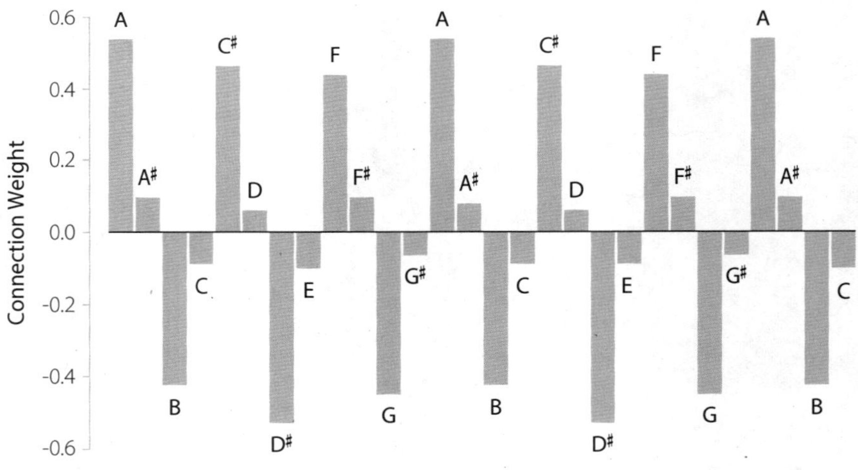

Figure 6-5 The connection weights from the 28 input units for pitch to Hidden Unit 3 in the network trained to classify triad types for different keys and inversions.

The final pattern of connectivity to consider involves Hidden Unit 4 (Figure 6-6). This pattern demonstrates octave equivalence, with the exception of the final weight for C, which is a dramatic outlier. (This particular note is included in only one chord; occasionally I find that networks can solve problems by treating rarely used input units uniquely.) Paying attention only to the sign of connection weight (and ignoring the deviant final weight!), it is notable that connection weight sign alternates from positive to negative every semitone. This divides the pitches into groups that belong to different interval cycles of major seconds. One of these cycles contains A, B, C♯, D♯, F, and G; the other cycle contains A♯, C, D, E, F♯, and G♯. This pattern of connection weights also demonstrates tritone equivalence; pitches separated by a tritone have roughly the same weight.

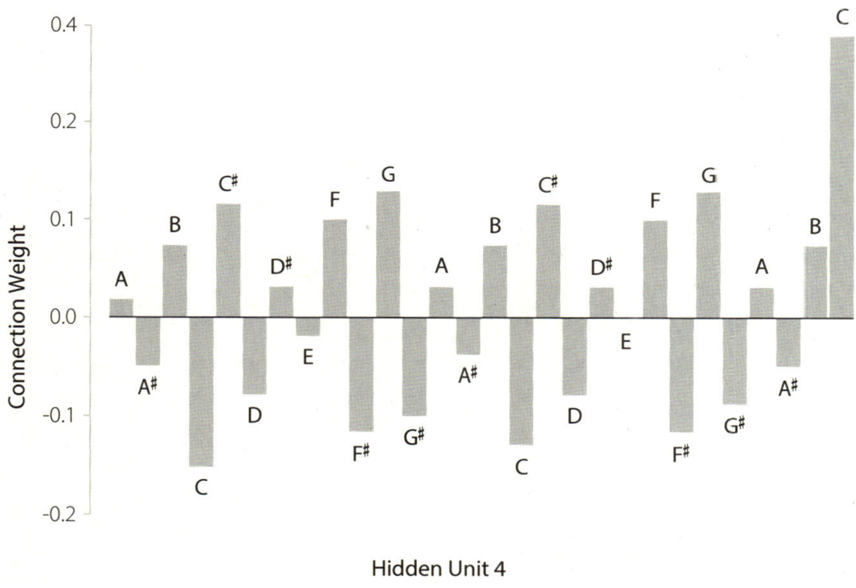

Figure 6-6 The connection weights from the 28 input units for pitch to Hidden Unit 4 in the network trained to classify triad types for different keys and inversions.

The pattern of connection weights that feeds into each of this multilayer perceptron's hidden units clearly indicates that pitches and pitch-classes are often organized into distinct equivalence classes that involve musical intervals. We will see below that this is a common discovery in networks of value units trained to make judgments about chords. Network interpretation can often be implified by seeking these sorts of relationships out, but they are even more interesting because they point to a very different kind of formal music theory.

To get a better understanding of these relationships and their implications, let us first detail the properties of various interval cycles. I will then consider a second multilayer perceptron trained on a different chord classification task, and discover interesting equivalence classes between pitches and pitch-classes when we examine its hidden unit weights.

6.3 Interval Cycles and Strange Circles

Section 6.2 described a multilayer perceptron for classifying triad types. The connection weights feeding into its hidden units grouped different sets of pitches and pitch-classes into various equivalence classes. I call such equivalence classes strange circles. They are circles, in the sense that they involve sets of pitch-classes that belong to musical sets called interval cycles, each of which can be represented as a circle of pitch-classes. They are strange, although, as far as networks are concerned, pitch-classes that belong to the same interval cycle are all treated as being the identical note. In other words, when I find networks that employ strange circles in their internal representations, these networks are using fewer than the 12 pitch-classes that are the core components of Western music theory.

I now explore such relationships in more detail, because they appear frequently when I train artificial neural networks on tasks involving musical harmony (Yaremchuk & Dawson, 2005, 2008). I begin by using the notion of interval cycles from music theory (Roig-Francolí, 2008) to generate the possible interval equivalences that might be discovered inside a network. I will generate a catalogue of the various interval cycles that can be defined for Western music, and then point out that each of these cycles could be used as a strange circle. Later in this chapter, and in chapters that follow, we will see several examples of networks whose internal structure reveals such strange circles.

6.3.1 Piano Geography

Our general approach to defining strange circles is to generate geometric representations of interval cycles. The foundation of our geometric representation of pitch-class relationships is a physical artifact, the piano keyboard. Each piano key, when struck, produces a unique pitch. For instance, the lowest (left-most) shaded note at the top of Figure 6-7 corresponds to the pitch "middle C," which is sometimes designated with the name C4.

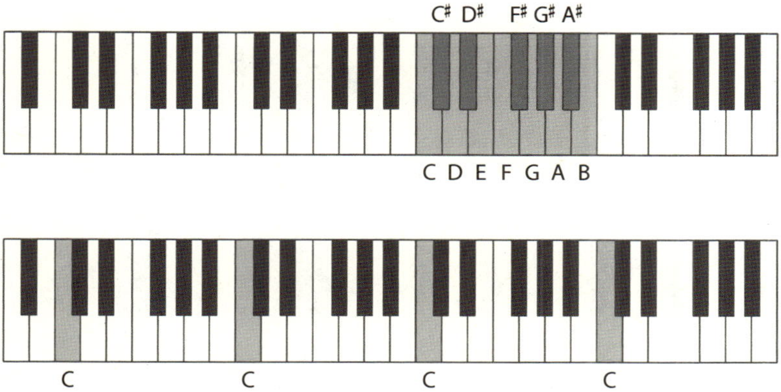

Figure 6-7 The geography of the piano.

The layout of piano keys is quite regular. This is evident in Figure 6-7 from the arrangement of black keys, which alternate in groups of twos and threes across the figure. The pattern of 12 differently named piano keys at the top of Figure 6-7 repeats itself along the keyboard.

While every piano key plays a differently pitched note, that note belongs to one of the 12 pitch-classes of Western music that we have already encountered. Therefore, several different piano keys play different pitches that all belong to the same pitch-class, and they occur at regular intervals along the piano keyboard. This is illustrated at the bottom of Figure 6-7, which highlights the locations of four different instances of the pitch-class C. Nearest neighbours on the keyboard that belong to the same pitch-class are separated by a span of 12 adjacent piano keys. This distance is equivalent to 12 semitones, or a musical interval of a perfect octave.

6.3.2 Distance and Intervals

With respect to our piano geography, what is the distance between two notes? For instance, what is the distance between the highlighted notes C and E at the top of Figure 6-8? We can measure this distance in terms of the number of piano keys that separate the two notes.

Examine the top illustration in Figure 6-8. If one starts at the highlighted C and moves up in pitch (i.e., to the right along the keyboard), then the first key encountered is C♯, the second is D, the third is D♯, and the fourth is E. Therefore, the distance between C and E is four piano keys. Alternatively, we can say that the distance between C and E in this figure is four semitones, which is a musical interval of a major third.

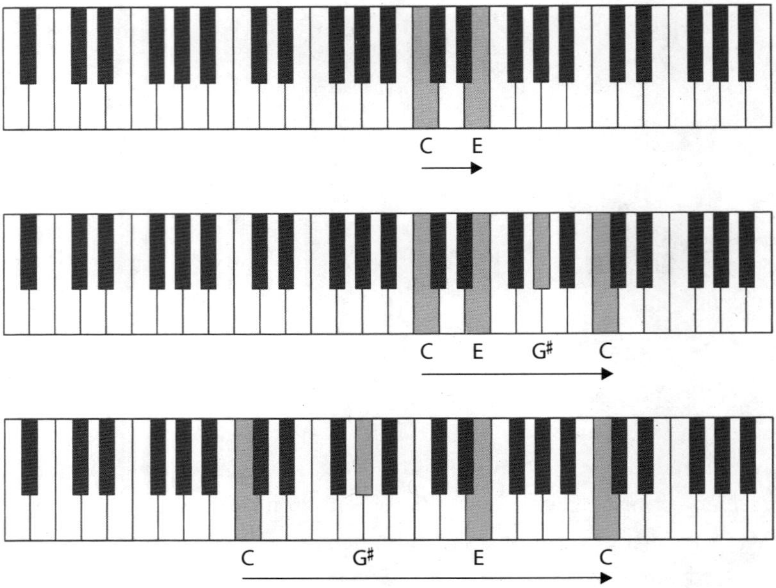

Figure 6-8 Using the number of piano keys as a measure of the distance between pitches.

We can identify sets of pitches that are spaced the same distance apart by continuing to count the same number of keys to the next note, as illustrated in the middle part of Figure 6-8. The distance up from C to E is four piano keys; if we move the same distance up from E, we reach G♯. If we move up four piano keys from G♯, we reach another C. In other words, if we start at C, and always move four piano keys up, we will only encounter three different pitch-classes: C, E, and G♯.

It is convenient to arrange a set of pitch-classes picked out by moving a fixed distance along the piano in a circle. The circle that results when a distance of four piano keys is used starts with C, moves next to E, moves next to G♯, and then returns to C. It is the fact that after a few moves we return to the pitch-class that we started from (C in this example) that motivates arranging this set of pitch-classes in a circle. We literally come full circle back to the pitch-class from which we started. Such circles of intervals are called interval cycles (Roig-Francolí, 2008).

If we use the same distance between notes but start at a different piano key, we can define a different instance of the same interval cycle. For instance, if we start at C♯ and move up four keys at a time, our circle will only include the pitch-classes

C♯, F, and A. With a between-note distance of four piano keys, we can define four different circles of three pitch-classes.

Interestingly, we can encounter the same subset of pitch-classes by starting at C and counting a different distance. The bottom part of Figure 6-8 shows that G♯ is eight piano keys higher than C, and that E is eight piano keys higher than G♯. With this different distance, we encounter the same pitch-classes highlighted in the middle of Figure 6-8, but encounter them in a different order. That the same subset of pitch-classes are encountered when one moves different distances along the keyboard indicates that we can consider these pitch-classes as being separated by two different musical intervals (Roig-Francolí, 2008). For instance, C and E can be considered to be a major third apart (four piano keys) or a minor sixth apart (eight piano keys).

Clearly, there is a musical interpretation for each distance between pitches measured in terms of number of piano keys or in terms of semitones. Table 6-1 presents the names of the different intervals in Western music, as well as their associated distance between pitches. Note that this table indicates that there are only 13 different types of interval cycles to be concerned about, because there are only 13 different semitone distances listed in the table, ranging from 0 semitones to 12 semitones. The next sections consider the properties of these possible interval cycles.

Table 6-1 The 13 possible distances between pitches that can be used to create interval cycles.

Distance between pitches in semitones	Interval name
0	Perfect unison
1	Minor second
2	Major second
3	Minor third
4	Major third
5	Perfect fourth
6	Tritone
7	Perfect fifth
8	Minor sixth
9	Major sixth

Distance between pitches in semitones	Interval name
10	Minor seventh
11	Major seventh
12	Perfect octave

6.3.3 Single Interval Cycles

The first set of interval cycles to discuss are those distances between pitches which, when used, combine all 12 different pitch-classes into a single set or a single interval cycle. To begin, let us consider starting on a piano keyboard at a note named C, and moving up from this note (i.e., to the right along the keyboard) a distance of one piano key. The first note that we will encounter is C♯. Moving the same distance up from C♯ we will next encounter D. Continuing to move along the keyboard in this fashion, we will encounter every possible pitch-class before we finally reach another note named C. The set of pitch-classes that we encounter, and the order in which we encounter them, can be represented in a single circle (Figure 6-9). (To make this figure easier to compare to other versions, instead of drawing the circle, we arrange the pitch-classes in a circle, and then draw a radius to the centre of the circle to place each pitch-class on a "spoke.") Because the distance between adjacent notes in this circle is one piano key (one semitone), the interval between adjacent notes is a minor second. Therefore, we can name Figure 6-9 the circle of minor seconds. Note that moving in a clockwise direction around this circle is equivalent to moving up (i.e., to the right) along a piano keyboard, and moving in a counter-clockwise direction is equivalent to moving down along a piano keyboard.

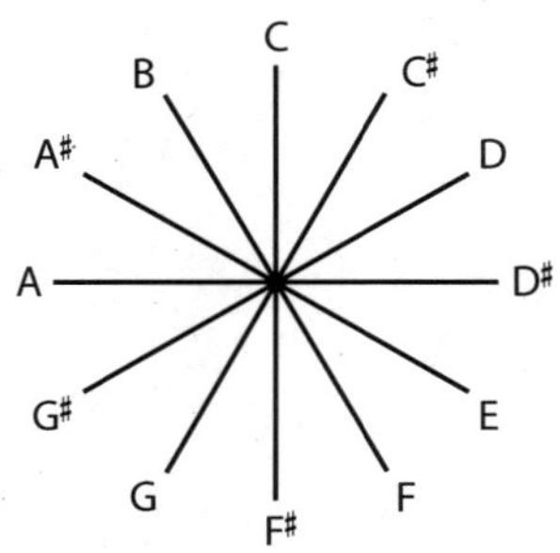

Figure 6-9 The circle of minor seconds.

Earlier we noted that by choosing a different distance to move along the piano keyboard we will encounter the same set of pitch-classes, but will do so in a different order. This is the case if we start at a note named C and move up the keyboard a distance of 11 keys (or 11 semitones), which corresponds to a musical interval of a major seventh. The order of notes that are encountered is illustrated in Figure 6-10 as a circle of major sevenths.

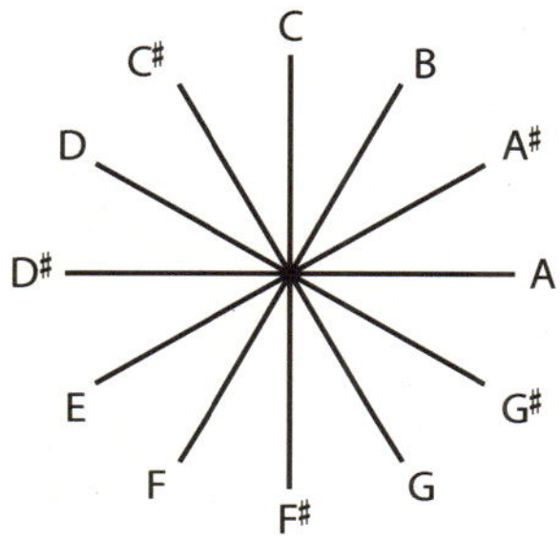

Figure 6-10 The circle of major sevenths.

An inspection of the circle of major sevenths indicates that it picks out the same pitch-classes as does the circle of minor seconds, but it does so in a different order. Indeed, the two circles are complementary: the circle of major sevenths is a mirror reflection of the circle of minor seconds.

We have seen that distances of one or 11 piano keys pick out all 12 pitch-classes; they can be arranged around a single circle to represent the order in which pitch-classes are encountered. Another distance that picks out all 12 pitch-classes is one of five piano keys (or semitones). Moving up a piano keyboard at this distance produces the circle of perfect fourths, illustrated in Figure 6-11. Note that this circle arranges pitch-classes in a very different order than we saw in Figures 6-9 and 6-10.

As was the case with the circle of minor seconds, the circle of perfect fourths has a complementary circle that is its reflection. The circle of perfect fifths is produced when one starts at C and moves up a piano keyboard a distance of seven piano keys or seven semitones (Figure 6-12). Note that if one were to move counter-clockwise around the circle of perfect fourths, one would encounter pitch-classes in the same order as moving clockwise around the circle of perfect fifths. This is because when a musical interval of a perfect fourth is inverted, the result is a musical interval of a perfect fifth.

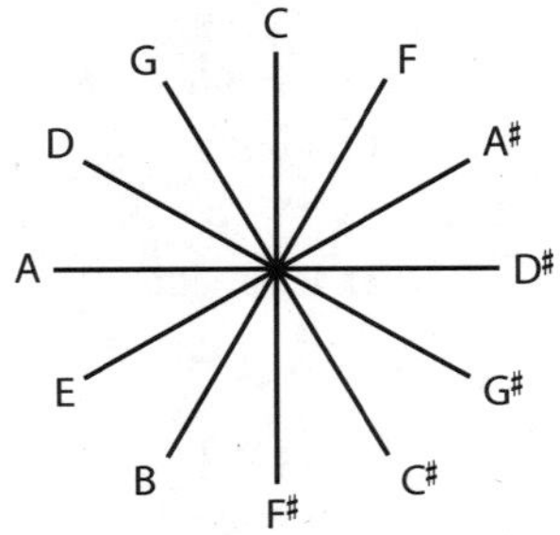

Figure 6-11 The circle of perfect fourths.

Of the four interval cycles presented in this section, the one most frequently seen in music is the circle of perfect fifths. Music students learn how to use this circle to determine the number of sharps or flats are in a particular musical key. Later we will see how this circle serves as a map to guide a musician who wishes to play the chords of a particular jazz progression, the ii-V-I. The circle of minor seconds is also commonly encountered; for instance, it is frequently found in geometric discussions of musical regularities (Tymoczko, 2011).

Each of the interval cycles presented in this section is defined musically: each is created by moving upward along a piano keyboard in steps of a set distance. However, these cycles—as well as those presented in following sections—can also be considered as mathematical objects called manifolds.

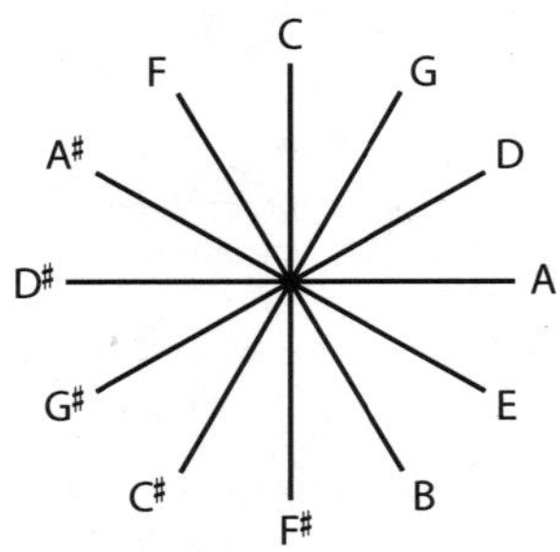

Figure 6-12 The circle of perfect fifths.

A manifold is a surface upon which objects are represented as points. Manifolds have specific shapes, and exist in a space that has a set number of dimensions. For instance, all of the manifolds described in this section are circular shapes. They are one-dimensional manifolds, in the sense that you can trace the entire shape with a finger without having to lift the finger up before the trace is completed. These manifolds are embedded in a higher-dimensional space. For instance, each of these manifolds is depicted in a two-dimensional space, the plane of the page in which each figure is drawn. In short, each of the circles that we have discussed is a one-dimensional manifold, circular in shape, embedded in a two-dimensional space.

The shape of a manifold is important, because it constrains how one moves from location to location along the manifold's surface. That is, to move from one location to another, one must never leave the manifold's surface. This property of manifolds has been used in theories of visual perception and visual imagery (Farrell & Shepard, 1981; Shepard, 1984) to model the appearance of three-dimensional objects as they move or as they are mentally rotated.

In describing the single interval cycles as manifolds, I am asserting that their shape and layout place certain constraints on how one can move from one note (i.e., from one point on the manifold's surface) to another. For instance, on the circle of perfect fifths, to move from C to D one must necessarily pass through an intermediate location, G. This is because G occupies an intermediate location between C and D on the manifold's surface.

Interpreting each of the circles as a manifold has further implications concerning the notion of distances between notes (Tymoczko, 2011). Each manifold is derived by moving along a piano keyboard at a set distance. However, after creating a manifold, I could measure the distance between notes along the manifold's surface. For instance, in the circle of perfect fifths, G is one unit of distance away from C because it is next to C on the manifold; similarly, D is two units away from C on the same manifold.

From this perspective, the distance between notes depends upon a particular context: the specific manifold under consideration. In the circle of perfect fifths, G is one distance unit away from C. In the circle of minor seconds, the shortest distance between C and G is five distance units. The idea that the distance between notes can be measured along a manifold, and that the size of this distance depends on the particular manifold being considered, is strongly related to Dmitri Tymoczko's idea of measuring distance between notes in the context of different musical scales (Tymoczko, 2011).

Importantly, if one uses a musical manifold or a musical scale as the context in which to measure the distance between pitch-classes, then one is tacitly assuming that different points on the manifold represent different pitch-classes. Of particular interest to us in this chapter is that in some instances artificial neural networks capture pitch-classes that can be represented in a manifold, but do not treat these pitch-classes as being different. For these networks, the manifold is an equivalence class, because all the pitch-classes that belong to it are the same. However, in order for equivalence classes of this sort to be useful there must be more than one, so that some pitch-classes belong to one equivalence class but not others. In the next section, we consider circles of musical intervals that pick out different and complementary subsets of pitch-classes.

6.3.4 Pairs of Interval Cycles

In the previous section, I detailed four different musical manifolds that are interval cycles. They are single in the sense that one manifold captures all 12 pitch-classes on its surface. Next, I will describe some new manifolds, each of which only captures half of the available pitch-classes. As a result, two different versions of the same process—moving along the piano keys—are required to build two complementary manifolds which, when combined, capture all 12 pitch-classes.

Consider starting at a C note on a piano, and moving upward a distance of two keys to the next note, which will be D. Following this procedure, we will next encounter E, F♯, G♯, and A♯ before encountering another C. One can build a manifold of these notes —a circle of major seconds—but it will only hold six of the 12 available pitch-classes. If one repeats this process, but start on C♯ instead of C, we will create a second circle of major seconds that complements the first because it captures the remaining six pitch-classes. Figure 6-13 illustrates these two circles of major seconds.

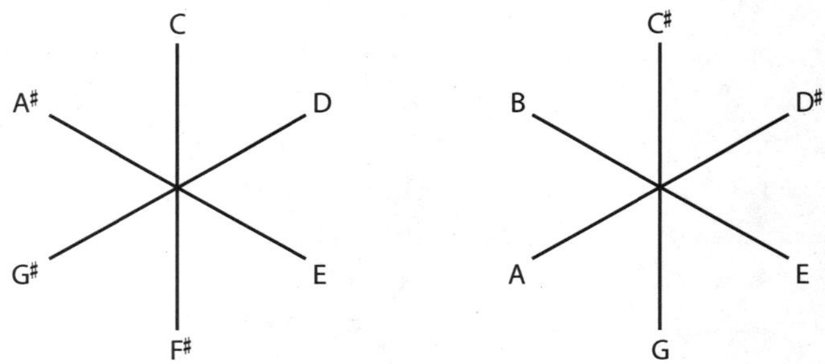

Figure 6-13 The two circles of major seconds.

If one takes a major second interval and inverts it, then the result is a minor seventh. This is because after inverting the major second interval, the distance between the low D and the higher C would be 10 semitones. From this, one should expect to be able to produce two circles of minor sevenths that complement the circles of major seconds illustrated in Figure 6-13. Indeed, if one gains at C and moves upward along the piano keyboard 10 keys at a time a circle of minor sevenths that is a reflection of the first circle of major seconds in Figure 6-13 is produced. Starting instead at C♯, one produces a manifold that complements the first circle of minor sevenths and reflects the second circle of major seconds in Figure 6-13. Figure 6-14 provides these two circles of minor sevenths.

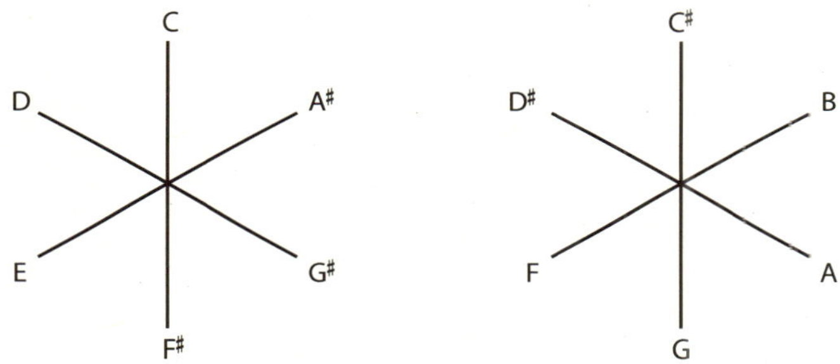

Figure 6-14 The two circles of minor sevenths.

6.3.5 Trios of Interval Cycles

In this section, I define sets of three complementary manifolds, each of which captures four pitch-classes; all three combined contain all 12 pitch-classes. Imagine starting on the piano at some C note and moving upward along the keyboard a distance of three keys. The first note encountered is D♯. Moving the same distance upward, one encounters F♯, then A, and then another C. Thus, this defines a circle that captures four of the 12 pitch-classes. This manifold is a circle of minor thirds, because if two adjacent notes are three semitones apart, they are separated by an interval of a minor third.

To define other, complementary, manifolds I must move the same distance along the keyboard but from different starting points. If we start at C♯ instead of C, a second circle of minor thirds is defined; if we start at D instead of C, a third circle of minor thirds is defined. Figure 6-15 illustrates the three possible circles of minor thirds.

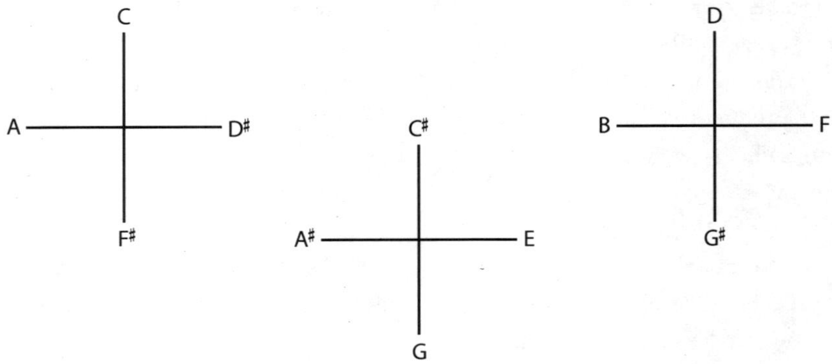

Figure 6-15 The three circles of minor thirds.

If one takes a minor third and inverts it, then one produces an interval of a major sixth whose notes are nine semitones apart. Not surprisingly, if I start at each of the three notes used above (C, C♯, D) and move along the piano nine keys at a time, I produce three different complementary circles of major sixths, shown in Figure 6-16. Each of these circles is a reflection of one of the circles of minor thirds illustrated in Figure 6-15.

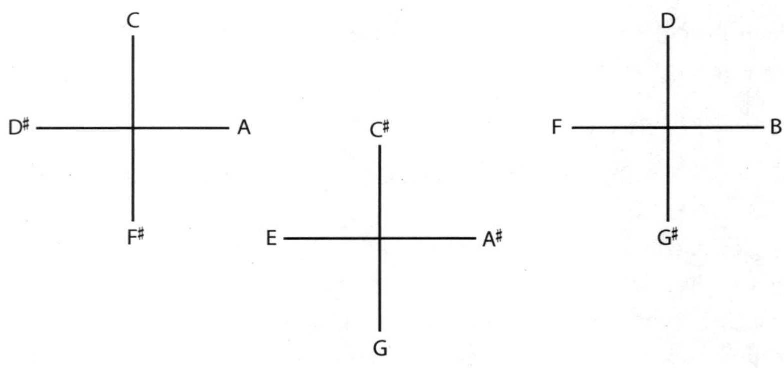

Figure 6-16 The three circles of major sixths.

6.3.6 Quartets of Interval Cycles

Imagine starting on the piano at some C note and moving upward along the keyboard a distance of four keys. The first note encountered is E. Moving the same distance upward, one encounters G♯, and then encounters another C. Thus, this defines a circle that captures only three of the 12 pitch-classes. This manifold is a circle of major thirds.

Three other circles of major thirds are possible, and are required to capture the remaining pitch-classes. I create them by moving the same distance along the piano keyboard, but from different starting points: C♯, D, and D♯ respectively. Figure 6-17 illustrates the four circles of major thirds.

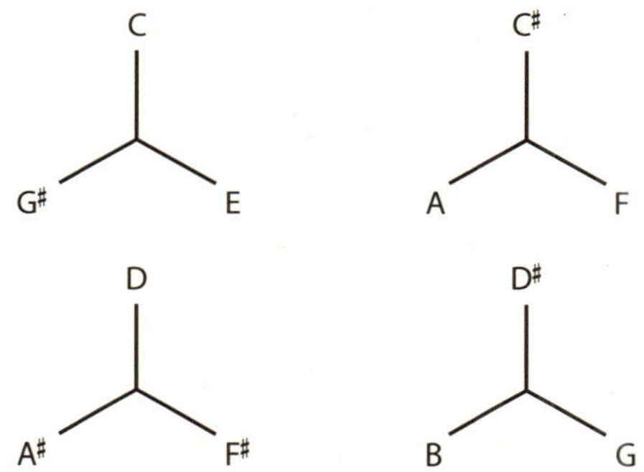

Figure 6-17 The four circles of major thirds.

If one takes a major third and inverts it, one produces an interval of a minor sixth whose notes are separated by eight semitones. If one starts at each of the four starting points used to create the manifolds of Figure 6-17 (C, C♯, D, D♯) and moves along the piano eight keys at a time, then one produces the four complementary circles of minor sixths, each of which is illustrated in Figure 6-18. Each of these circles is a reflection of one of the circles illustrated in Figure 6-17.

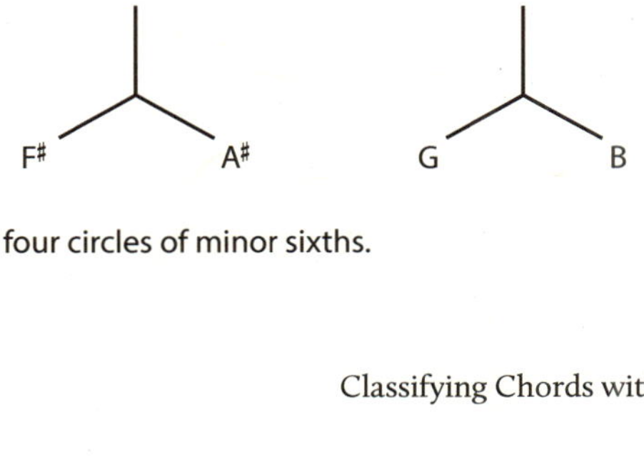

Figure 6-18 The four circles of minor sixths.

6.3.7 Sextets of Interval Cycles

If one starts at some C note on the piano keyboard and moves up six piano keys, one encounters F♯. Moving up another six piano keys one reaches another C. This defines a simple manifold that contains only two points. To capture the remaining pitch-classes requires starting from five additional notes on the keyboards. This produces the six different circles of tritones that Figure 6-19 provides.

Figure 6-19 The six circles of tritones.

The inversion of a tritone is itself a tritone, because this interval is defined by six semitones, a distance that is exactly half an octave. Thus, no other interval cycles are reflections of those illustrated in Figure 6-19.

6.3.8 Dodecal Interval Cycles

Choose some note C on a piano, and move 12 keys—a perfect octave—upward. You reach another C. This produces a special case manifold, a "circle" that represents a single pitch-class as a single point. Obviously 11 other such manifolds are required to capture the remaining pitch-classes (Figure 6-20).

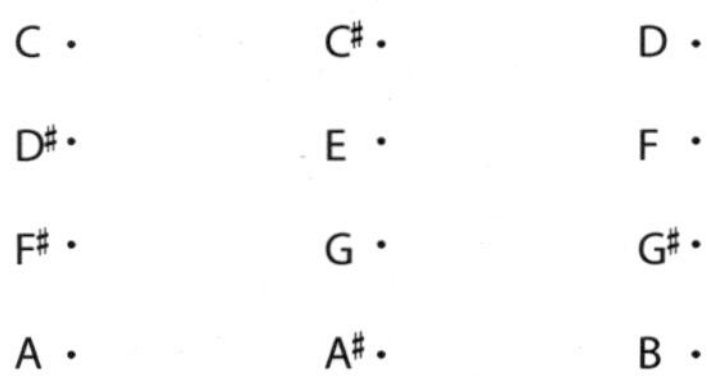

Figure 6-20 The 12 circles of octaves, or circles of unison.

If one inverts an octave interval by raising the lower note an octave, the result is two identical notes—the distance between them is zero semitones. This musical interval, perfect unison, produces exactly the same set of 12 manifolds given in Figure 6-20.

6.3.9 Strange Circles

We introduced single interval cycles in Section 6.3.3 (e.g., the circle of minor seconds and the circle of perfect fifths). Then we discussed a number of interval cycles in which more than one cycle existed for the same interval. These were the two circles of major seconds, the two circles of minor sevenths, the three circles of minor thirds, the three circles of major sixths, the four circles of major thirds, the four circles of minor sixths, the six circles of tritones, the 12 circles of perfect octaves and the 12 circles of unison.

As was the case for the single circles of intervals, each of these multiple circles is a manifold. For instance, on one of the circles of major seconds the distance between C and D is one unit. However, we can interpret each of these manifolds in a different way: as an equivalence class. For instance, let us return for a moment to consider the hidden units of the multilayer perceptron that learned to classify triads (Section 6.2). Hidden Unit 1 (Figure 6-3) organizes inputs in terms of circles of minor thirds: all the pitches that belong to the same circle have the identical weight from the input unit to this hidden unit. However, this also means that all of the different (to us) members of this circle are identical for this hidden unit. Similarly, Hidden Unit 2 (Figure 6-4) assigns the same connection weights to inputs that belong to the same circle of major thirds; Hidden Unit 3 (Figure 6-5) also organizes inputs into equivalence classes based on circles of major thirds. Hidden Unit 4 (Figure 6-6) organizes inputs into equivalence classes based on circles of major seconds (if one only examines connection weight signs) and into equivalence classes based on circles of tritones (if one examines both the magnitude and sign of each connection weight).

I call equivalence classes based upon circles of intervals strange circles. These circles are strange in two ways.

First, while entities like the two circles of major seconds or the three circles of minor thirds are proper components of music theory, they are rarely encountered in theories of tonal music—music associated with a tonal centre. Instead, they are more likely to be encountered in theories about atonal or post-tonal music (Laitz, 2008; Roig-Francolí, 2008; Straus, 2005). This is because these interval cycles are all examples of symmetric sets of pitch-classes. That is, one can draw at least one axis through images such as those illustrated in Figures 6-13, 6-15, 6-17, and 6-19 such that the arrangement of the pitch-classes on one side of the axis mirrors the arrangement of those on the other side.

Symmetric sets of pitch-classes are important elements in post-tonal music. This is because "the basis of tonality's gravitational field, which pulls scale degrees and harmonies toward tonic, is predicated on asymmetry" (Laitz, 2008, p. 812). For instance, key elements of tonal music, like the major and harmonic minor scales that we have encountered earlier, depend upon the asymmetric arrangement of pitch-classes that result when one creates a scale in which neighbouring notes are spaced apart by an irregular arrangement of tones and semitones (see Chapter 2). If this asymmetry is eliminated by choosing sets of pitch-classes that are equally spaced (as in the strange circles), then "a sense of goal-directed motion and tonal grounding disappears because every scale step is as stable (or as unstable) as every other step" (Laitz, 2008, p. 813).

In short, when one trains a network to make a musical judgment about tonal musical stimuli, and discovers that it does so by employing symmetric interval cycles, then this is indeed strange.

A second reason for calling these circles strange is that when one finds them in networks, they typically involve assigning different input units (i.e., pitch-classes) identical connection weights that feed into the same hidden unit. This means that as far as this hidden unit is concerned, these different (to us) pitch-classes are identical. In other words, a hidden unit that assigns one connection weight to the pitch-classes that belong to one circle of major seconds, and assigns another connection weight to all of the pitch-classes that belong to the other circle of major seconds, is operating as if music is constructed from only two pitch-classes instead of 12.

On the one hand, the use of equivalence classes to represent musical regularities is not odd. For instance, we have already encountered the notion of octave equivalence that motivated our discussion of pitch-class representations in Chapter 2. We saw in Chapter 3 that scales constructed on different tonics could be assigned to equivalence classes based on scale mode (e.g., major vs. harmonic minor). Clearly, the use of equivalence classes is central to music theory.

On the other hand, the foundation of almost all theory concerning Western music assumes the existence of 12 different pitch-classes. For instance, when the principle of octave equivalence is invoked in the theory of atonal or post-tonal music (Forte, 1973; Roig-Francolí, 2008; Straus, 2005), this implies the assumption of 12 different pitch-classes. Music theory has not explored the consequences of using interval cycles to define equivalence classes that imply fewer than 12 pitch-classes.

Let us now turn to another network whose interpretation reveals that it treats a number of different pitch-classes as being the same, because they have the same connection weight. Whenever this occurs, one finds that the equivalence class that they belong to is one of the strange circles described above.

6.4 Added Note Tetrachords

6.4.1 Tetrachords

The major and minor scales that serve as the foundation for much of Western music are rooted in musical formalisms invented by the ancient Greeks. The foundation of Greek music was not the scale but the tetrachord. The Greek tetrachord was a set of four different notes, the lowest always separated from the highest by an interval of a perfect fourth. Two additional notes were placed between these two, carving the tetrachord's perfect fourth into three smaller intervals. There were three main types of tetrachords, depending upon the choice of the inner two notes (Barbera, 1977; Chalmers, 1992). Our modern major and harmonic minor scales are constructed from two adjacent Greek tetrachords.

Figure 6-21 Added note tetrachords in the key of C major.

The modern definition of tetrachord includes a much wider variety of chords than does the Greek definition. A modern tetrachord is any chord that includes four different pitches. Figure 6-21 illustrates the construction of a subset of modern tetrachords. The top line of this score provides the notes of the C major scale. In the middle line, each of these notes serves as the root of a triad. The added notes are always two scale notes higher than the lower note, so triads are constructed by skipping over notes. For instance, the C major triad is C-E-G, which skips over D and F. Similarly, the D minor triad is D-F-A (skipping over E and G), and so on. This process produces three different major triads (C, F, and G), three different minor triads (Dm, Em, and Am), and one diminished triad (Bdim).

The last line in Figure 6-21 converts each triad into a tetrachord by adding another note from the C major scale. Again, the added note is two scale notes higher than the highest note in the triad. For instance, the Cmaj7 tetrachord is C-E-G-B (skipping over the A to add the B). Similarly, the Dmin7 tetrachord is D-F-A-C, and so on. Each of these tetrachords is a seventh chord. There are different types of these tetrachords created from this process: two major seventh chords (Cmaj7 and Fmaj7), three minor seventh chords (Dm7, Em7, and Am7), one dominant seventh chord (G7), and one minor seventh flat fifth chord (Bm7♭5).

The same approach to chord construction can be applied to any major scale, producing a set of seven different chords for each key. However, if I create these seven different chords for each of the 12 major keys, then I will not create 84 unique chords. This is because when a pitch-class representation is used the same chord will appear in different musical keys. For example, in the set of 84 chords, each min7 chord will appear three different times, and each maj7 chord will appear twice. As a result, our total set of 84 chords will include 48 unique tetrachords and 36 duplicates of some of these chords.

6.4.2 Tetrachord Properties

In order to illustrate networks that solve musical problems by assigning notes to equivalence classes based upon circles of intervals, we will consider a multilayer perceptron that is presented modern tetrachords of the type illustrated in Figure 6-21, and which learns to assign one to each of four different tetrachord classes. Prior to describing this network, let us consider the musical properties of these chords.

Earlier in this chapter, I noted that each different type of triad possesses a particular pattern of musical intervals between adjacent notes. The same is true for the different tetrachords. For instance, consider the Cmaj7 tetrachord in root position, whose notes (in order) are C, E, G, and B. There is an interval of a major third from C to E, of a minor third from E to G, and of a major third from G to B. This pattern of intervals distinguishes this type of tetrachord from the other three types, as can be seen from the third column of Table 6-2 below.

Of course, if one considers the distances between nonadjacent notes in a tetrachord, then there are more intervals than those presented in the third column of Table 6-2. In order to obtain a deeper understanding of the structure of these tetrachords we can use musical set theory (Forte, 1973) to determine each tetrachord's Forte number, prime form, and interval-class vector (ic vector). The final three columns of Table 6-2 provide the results of this analysis.

Table 6-2 Musical properties of each type of tetrachord in Figure 6-21.

Chord type	Example	Intervals between adjacent notes	Forte number	Prime form	IC vector
Major 7	[C, E, G, B]	major third - minor third - major third	4-20(12)	0,1,5,8,	101220
Minor 7	[D, F, A, C]	minor third - major third - minor third	4-26(12)	0,3,5,8,	012120
Dominant 7	[G, B, D, F]	major third - minor third - minor third	4-27	0,2,5,8,	012111
Minor 7flat5	[B, D, F, A]	minor third - minor third - major third	4-27	0,2,5,8,	012111

Note. The first column provides the chord type, and the second column provides an example of the chord. The third column provides the structure of the chord in terms of the musical intervals between adjacent pitches. The final three columns provide descriptors of the chord type taken from Forte's (1973) set theory, including Forte's classification number for each chord type, the prime form of the chord, and the IC vector that provides the interval structure of the chord.

Two particularly interesting findings emerge from this set-theoretic analysis. First, both the dominant seventh and the minor seventh flat fifth tetrachords have the same prime form and the same ic vector. This means that a network may only be able to differentiate these two tetrachords by considering the specific order in which the component musical intervals occur. Second, the ic vectors for each tetrachord type provide some indication of the musical regularities that a network may be able to exploit to differentiate tetrachord types, as detailed below.

In an ic vector, the first number indicates how many minor second/major seventh intervals occur in a musical object. The second number indicates the frequency of major second/minor seventh intervals. The third number indicates the frequency of minor third/major sixth intervals. The fourth number indicates the frequency of major third/minor sixth intervals. The fifth number indicates the frequency of perfect fourth/perfect fifth intervals. The sixth number indicates the frequency of tritones.

With this understanding of ic vectors, we can now see what the ic vectors in the final column of Table 6-2 reveal. For instance, a major seventh tetrachord is the only one that has a minor second or major seventh interval, and the only one that does not have a major second or a minor seventh interval in its structure. Neither the major nor the minor seventh tetrachords contain a tritone, but the other two types of chords do. The minor seventh tetrachord shares individual ic vector values with each of the other types of tetrachords; this means that it can only be distinguished from them by considering several interval types at the same time. For instance, it can be distinguished from the major seventh by the presence of a major second or minor seventh interval, but the other two types of tetrachords share this property. A minor seventh can only be distinguished from them by detecting the absence of a tritone interval.

Now let us turn to describing the training of a multilayer perceptron to detect these four different types of tetrachords, regardless of the musical key in which they occur.

6.5 Classifying Tetrachords

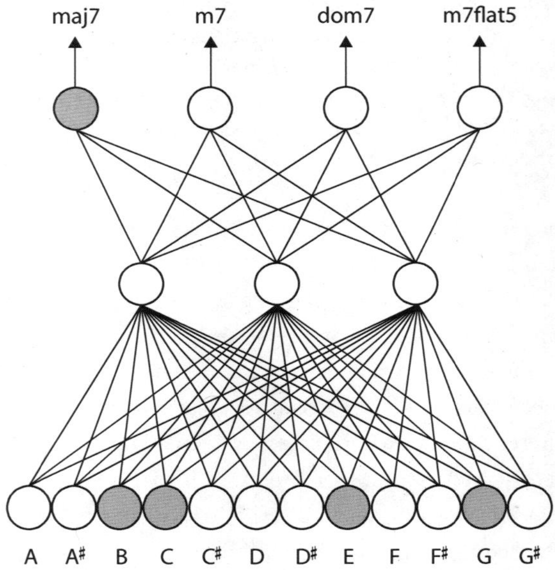

Figure 6-22 A multilayer perceptron that classifies tetrachords into four different types.

6.5.1 Task

Our goal is to train an artificial neural network, when presented with four notes that define a tetrachord constructed from the notes of a major scale (Figure 6-21), to identify the type of tetrachord (major seventh, minor seventh, dominant seventh, or minor seventh flat fifth), ignoring the key of the tetrachord.

At the end of training, this multilayer perceptron turns one output unit "on" to identify tetrachord type, and turns the remaining three output units "off," when presented a tetrachord. Thus, this network has four different output units, each one dedicated to representing a particular tetrachord type.

6.5.2 Network Architecture

Figure 6-22 presents the architecture that accomplishes this tetrachord classification task. It uses four output value units to represent tetrachord type, and requires three hidden value units to converge to a solution to this problem. It uses 12 input units to represent input tetrachords in the same pitch-class representation used for the training of the networks in several previous chapters. Figure 6-22 illustrates the presentation of the C major seventh tetrachord (grey input units), resulting in the "Major Seventh" output unit activating.

6.5.3 Training Set

The training set consists of 84 stimuli: the seven different tetrachords for a major scale (see Figure 6-21); these tetrachords are constructed for each of the 12 different major scales. Within this set of 84 stimuli there are 36 duplicate patterns (each maj7 tetrachord appears twice, and each min7 tetrachord appears three times). For the current network, this simply means that these two different types of tetrachords receive more training than the other two types. This difference is not relevant to the point that the network illustrates: the presence of strange circles in the connection weights of its hidden units. I encode each tetrachord as an input pattern in which four input units are activated with a value of one, and the remaining eight input units are all activated with a value of zero. Each input pattern is paired with an output pattern that requires one output unit to activate with a value of one, and the other three output units to activate with a value of zero. The output unit trained to activate is the one that represents the input pattern's correct tetrachord type.

6.5.4 Training

The multilayer perceptron is trained with the generalized delta rule developed for networks of value units (Dawson & Schopflocher, 1992) using the Rumelhart software program (Dawson, 2005). During a single epoch of training each pattern is presented to the network once; the order of pattern presentation is randomized before each epoch.

All connection weights in the network are set to random values between –0.1 and 0.1 before training begins. In the network to be described in detail below, each μ is initialized to zero but is then modified by training. A learning rate of 0.01 is employed. Training proceeds until the network generates a "hit" for every output unit for each of the 84 patterns in the training set. Again, I define a "hit" as activity of 0.9 or higher when the desired response is one or as activity of 0.1 or lower when the desired response is zero.

I explored a number of different network architectures with this training set. When networks have four or five hidden value units, the problem is very easy and is often solved in a few hundred epochs. However, in order to get a three-hidden-unit network to converge, each μ was modified during training. On some occasions, a three-hidden-unit network would converge very quickly. For instance, the network described in more detail in the next section converged after 11,566 epochs of training. However, on many occasions a three-hidden-unit network would settle to a local minimum and fail to converge to a solution even after more than 20,000 epochs of training. In other words, the network analyzed in the next section required a fair amount of patience during training!

Importantly, all of the networks trained on this problem developed patterns of connectivity that reflect equivalence classes defined by interval cycles. I simply focus on the smallest of these networks because with only three hidden units it is easier to consider some of its properties, such as its hidden unit space.

6.6 Interpreting the Tetrachord Network

6.6.1 Hidden Unit Space

How does this multilayer perceptron identify the four different types of tetrachords? Let us first consider the hidden unit space for this network, illustrated in Figure 6-23. This space is very sparse because the different instances of tetrachord types are very near one another in the space. Indeed, in many cases different tetrachords occupy the same coordinates in this space, which is why it appears to have so few symbols illustrated, even when there are 84 different input patterns plotted in this figure.

We saw such overlapping of points in an earlier hidden unit space, the one illustrated in Figure 4-2. It is important to realize that this feature of the hidden unit space is one of the key properties made explicit by this visualization of the hidden unit space. While the overlapping of symbols in Figure 6-23 seems to make the graph harder to inspect, it delivers a fundamental characteristic: as far as this network is concerned, chords that are of the same type, but which belong to different keys, are identical. This is why the different chords occupy the same location in the hidden unit space.

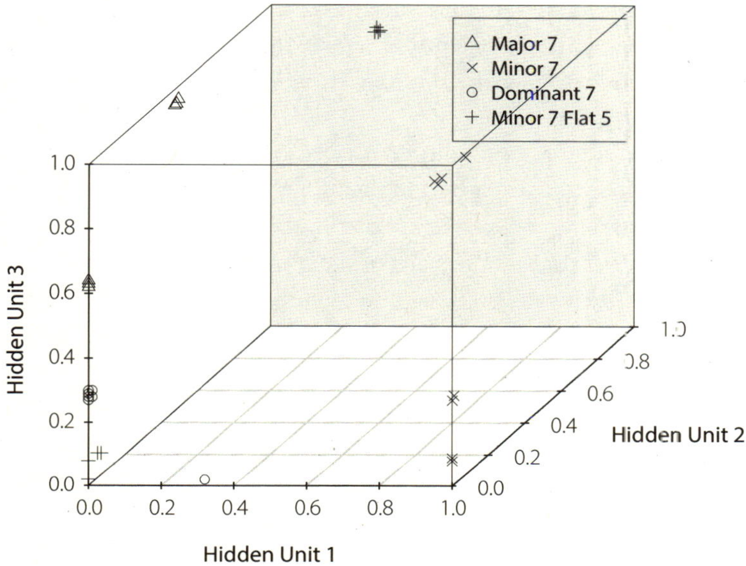

Figure 6-23 The hidden unit space for a multilayer perceptron trained to identify the four types of tetrachords.

In this space, all of the major seventh tetrachords are at two general locations at the cube's upper left. All of the minor seventh tetrachords fall along the front right edge of the Figure 6-23 cube. All of the dominant seventh tetrachords fall in two regions in the lower left-hand corner of the cube. All of the minor seventh flat fifth tetrachords fall either in a single tight area located in the upper back region of the cube or in a similar location in the lower left-hand corner at the front of the cube. Importantly, all of these locations of tetrachord types in the hidden unit space can easily be separated from the other tetrachords by two parallel planes that are carved through the space by each output value unit. Let us consider the tetrachord properties detected by each hidden unit.

The fact that all of the different types of tetrachords are near one another, typically in two different areas of the hidden unit space, suggests that each type of tetrachord produces a small number of different patterns of activity in the hidden units. We can confirm this by taking each type of tetrachord and examining the hidden unit activities produced by each. Three of the tetrachord types produced two distinct patterns of hidden unit activity, while the fourth (the minor sevenths) produced three distinct patterns of hidden unit activity. Table 6-3 presents the activity in each hidden unit, averaged over all of the hidden units that fall together in a particular type. Each row of hidden unit activities represents the general coordinates of tetrachord locations in Figure 6-23, and confirms our visual inspection of that figure.

Table 6-3 indicates that each subtype of a tetrachord shares some similarities in terms of some hidden unit activities, but they differ from each other in terms of the other activities that they produce. For instance, consider the three different subtypes of the minor seventh tetrachords. Each of these subtypes is similar in producing very high activity in Hidden Unit 1, and in producing very low activity in Hidden Unit 2. The three differ in terms of the activity that each produces in Hidden Unit 3: one produces very high activity in this unit, another produces near zero activity, and the third produces weak activity.

Table 6-3 The different patterns of hidden unit activity produced by different subsets of each type of tetrachord.

Chord type	Pattern	# of tetrachords	H1	H2	H3
Major 7	1	16	0.00	0.00	0.63
	2	8	0.00	0.48	0.95
Minor 7	1	12	0.96	0.04	0.95
	2	12	1.00	0.00	0.09
	3	12	1.00	0.01	0.27
Dominant 7	1	8	0.00	0.00	0.29
	2	4	0.32	0.00	0.02
Minor 7flat5	1	8	0.00	0.01	0.05
	2	4	0.32	0.95	0.94

Note. Each subset is given a number and the number of tetrachords that belong to that subset is indicated in the column labelled #. The H1, H2, and H3 columns provide the average activity produced in each hidden unit by a tetrachord that belongs to the subset.

In order to understand these different patterns of activity, and why particular tetrachords produce specific activities in specific hidden units, let us describe the pattern of connectivity between the 12 pitch-class input units and the three hidden units.

Once we have a sense of the regularities in these connection weights, we can use this knowledge to explain the regularities of Figure 6-23 and Table 6-3. In the sections that follow, we will consider hidden units from the most easily interpreted (musically) to the least; as a result, the order in which units are discussed will not agree with the names of the hidden units.

6.6.2 Hidden Unit 1

To begin, let us examine the pattern of connections between the 12 input units and Hidden Unit 1 (Figure 6-24). This figure provides a strong indication that this hidden unit classifies input pitch-classes in terms of the circles of tritones. That is, Figure 6-24 exhibits tritone equivalence: pitch-classes that are a tritone apart have essentially the same connection weight.

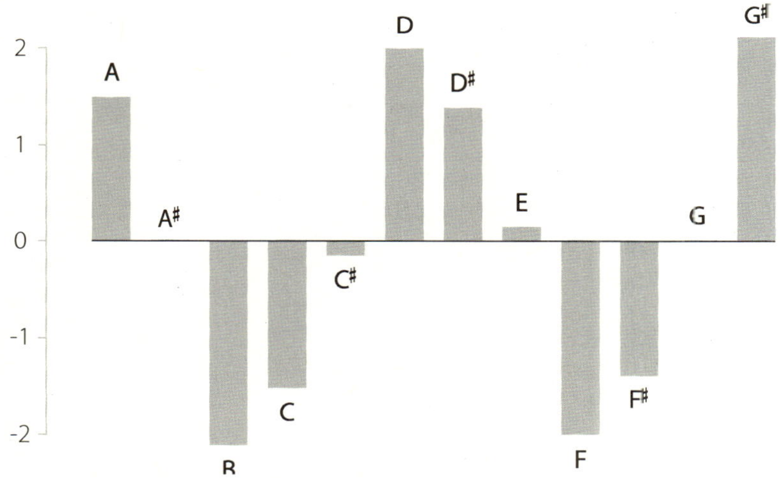

Figure 6-24 The connection weights from the 12 input units to Hidden Unit 1.

Tritone equivalence in this hidden unit is important because at the end of training its μ had a value of 0.00. We might expect that the presence of a pair of pitch-classes that are a tritone apart would produce an extreme net input, turning Hidden Unit 1 off. The ic vectors in Table 6-2 might lead us to predict that Hidden Unit 1 would therefore turn on to either major seventh or to minor seventh tetrachords (which do not include a tritone). However, the data in Table 6-3 does not support this prediction. Major seventh tetrachords never activate Hidden Unit 1. There must be something more sophisticated within the Hidden Unit 1 connection weights.

In networks described in earlier chapters, we observed a phenomenon called tritone balance. In tritone balance, pitch-classes a tritone apart had connection weights that were equal in magnitude but opposite in sign. As a result, if both pitch-classes were present they would cancel each other out. An examination of Figure 6-24 reveals that these connection weights are balanced, but not in terms of tritones. Instead, this hidden unit balances minor thirds. Pitch-classes that are a minor third apart have connection weights that are equal in magnitude but opposite in sign.

This relationship is explicit in Figure 6-25, which plots exactly the same connection weights from Figure 6-24 but stacks the weights from pitch-classes separated by a minor third on top of one another. The symmetry of this bar graph provides the evidence that these connection weights balance minor thirds.

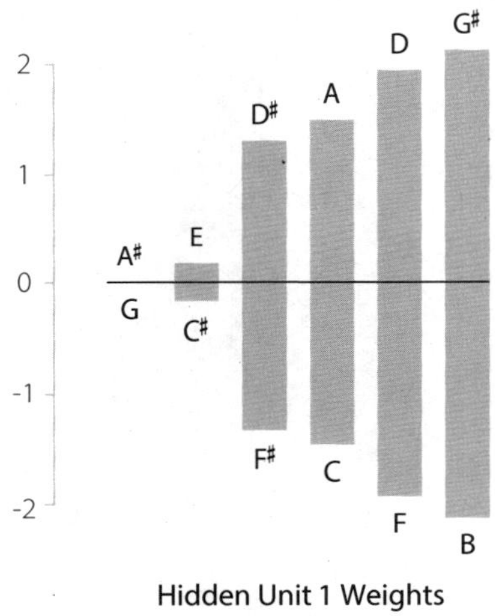

Figure 6-25 The connection weights of Figure 6-24 re-plotted so that weights from pitch-classes a minor third apart are stacked on top of one another.

Our earlier discussion of circles of minor thirds indicated that, unlike two pitch-classes separated by a tritone, one pitch-class is a minor third away from two other pitch-classes. For example, an examination of the first circle of minor thirds in Figure 6-15 indicates that C is not only a minor third away from A but is also a minor third away from D♯. If Hidden Unit 1 is truly balancing minor thirds, then we expect to find that one pitch-class is balanced with two others, and not just one.

This appears to be the case for Hidden Unit 1. Figure 6-26 presents yet another depiction of its connection weights from Figure 6-24. However, in this second figure each connection weight is stacked against the other pitch-class that is a minor third away (i.e., the pitch-class that it was not stacked against in Figure 6-25). Once again, this figure is very symmetrical, although the balancing is not as perfect as that shown in Figure 6-25. Together, Figures 6-25 and 6-26 reveal that Hidden Unit 1 balances a pitch-class with either of the pitch-classes that are a minor third away from it.

Our investigation of Hidden Unit 1 connection weights has revealed that they assign pitch-classes a tritone apart to the same equivalence class, and they balance minor thirds. Major seventh tetrachords do not include a tritone, but do include pitches that are a minor third apart (Table 6-2). Why, then, do these tetrachords fail to activate Hidden Unit 1? The answer to this question comes from the pitch-class representation used for the multilayer perceptron. This representation rearranges the structure of the various tetrachords, and the hidden units must process this rearranged structure.

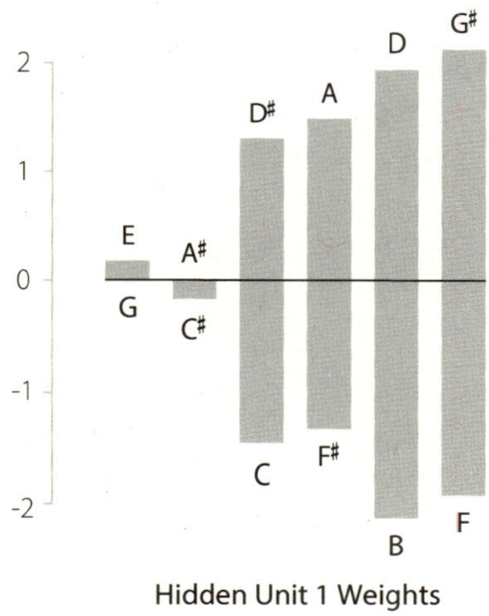

Figure 6-26 The connection weights of Figure 6-24 re-plotted so that weights from pitch-classes a minor third apart are stacked on top of one another. Note the difference in stacking between this figure and Figure 6-25.

Table 6-4 presents the pitch-class representation of two major seventh tetrachords in its first two rows; one is an example of Pattern 1 from Table 6-3 and the other is an example of Pattern 2. The key property to observe in both is that after being represented in this format, each contains two pitch-classes that are a minor second apart (A and A♯ in A♯maj7, D and D♯ in D♯maj7). This property is true of every major seventh tetrachord in the training set.

The presence of adjacent pitch-classes in any major seventh input pattern causes Hidden Unit 1 to turn off. This is because any four connection weights that include adjacent pitch-classes do not balance to produce a net input near zero to activate this unit. Instead, major seventh tetrachords that belong to Pattern 1 from Table 6-3 will include two balanced weights (e.g., D and F for A♯maj7), one near-zero weight (e.g., A♯ for A♯maj7), and one extreme weight that is out of balance with the other three (e.g., A for A♯maj7). Major seventh tetrachords that belong to Pattern 2 from Table 6-3 combine four weights that are even more unbalanced, causing more extreme net input (e.g., the D♯maj7 chord in Table 6-4).

Table 6-4 Example pitch-class representations of two major seventh tetrachords and three minor seventh tetrachords, along with the net input they provide to Hidden Unit 1 (Net) and its resulting activity.

Chord	A	A#	B	C	C#	D	D#	E	F	F#	G	G#	Net	H1 Activity
A#maj7	1	1	0	0	0	1	0	0	1	0	0	0	1.44	0.00
D#maj7	0	1	0	0	0	1	1	0	0	0	1	0	3.28	0.00
F#min7	1	0	0	0	1	0	0	1	0	1	0	0	0.12	0.95
Dmin7	1	0	0	1	0	1	0	0	1	0	0	0	0.00	1.00
Gmin7	0	1	0	0	0	1	0	0	1	0	1	0	0.00	1.00

Minor seventh tetrachords do not include adjacent pitch-classes in their pitch-class representation for the network, and as a result always include four connection weights that when combined produce near-zero net input to turn Hidden Unit 1 on. Table 6-4 also provides three example pitch-class representations of minor seventh tetrachords (one for each pattern in Table 6-3).

Each of the three minor seventh tetrachords presented in Table 6-4 contains two pairs of pitch-classes that are a minor third apart and have balanced weights; we can see all of these balanced pairs in Figures 6-25 and 6-26. For F♯min7 they are [A, F♯] and [C♯, E]. For Dmin7 they are [A, C] and [D, F]. For Gmin7 they are [A♯, G] and [D, F]. The balancing of each of these pairs of connection weights produces very small net inputs, and high Hidden Unit 1 activities, as presented in Table 6-4. Every other minor seventh tetrachord in the training set also exhibits this property. In other words, for minor seventh tetrachords Hidden Unit 1 behaves as expected from our interpretation of connection weights!

Let us now briefly turn to explaining the activity produced in Hidden Unit 1 by the two other types of tetrachords, the dominant seventh and the minor seventh flat fifth. Table 6-3 reveals that the majority of both types of these chords produce zero activity in Hidden Unit 1. This is consistent with our analysis of this unit's weights. Earlier, I noted that these weights exhibit tritone equivalence. Therefore, pitch-classes a tritone apart cannot cancel each other's signal out, because both pitch-classes are associated with the same connection weight. Indeed, in the pitch-class representation of each of the dominant seventh and the minor seventh flat fifth chords that turns this unit off, one finds two pitch-classes a tritone apart. Their combined weights produce an extreme net input that is very far from 1.

What is surprising about Table 6-3 is that a minority of both of these types of tetrachords produce weak activity in Hidden Unit 1. How is this possible if these stimuli include a tritone?

Table 6-5 presents the pitch-class representation of four example tetrachords that produce this surprising behaviour in Hidden Unit 1. All four of these chords include a pair of tones a tritone apart: [A, D♯] in B7, [C, F♯] in G♯7, [D, G♯] in Dmin7flat5, and [B, F] in Bmin7flat5. However, other intervals, when present, moderate the effect of the unbalanced tritone. For example, B7 includes the pitch-classes [A, F♯]; these provide a balanced minor third (Figure 6-26). The same is true for the pitch-classes [C, D♯] in C♯7, [F, G♯] for Dmin7flat5, and [D, B] in Bmin7flat5. The remaining two connection weights together produce less extreme net input (see Table 6-5) that produces moderate Hidden Unit 1 activity. In other words, for this subset of tetrachords, Hidden Unit 1 compromises its activity because it detects one interval that should turn it off (a tritone), but another that should turn it on (a minor third).

Table 6-5 Example pitch-class representations of two dominant seventh tetrachords and two minor seventh flat five tetrachords, along with the net input they provide to Hidden Unit 1 (Net) and its resulting activity.

Chord	A	A#	B	C	C#	D	D#	E	F	F#	G	G#	Net	H1 Activity
B7	1	0	1	0	0	0	1	0	0	1	0	0	−0.51	0.32
G#7	0	0	0	1	0	0	1	0	0	1	0	1	0.60	0.32
Dmin7flat5	0	0	0	1	0	1	0	0	1	0	0	1	0.60	0.32
Bmin7flat5	1	0	1	0	0	1	0	0	1	0	0	0	−0.51	0.32

6.6.3 Hidden Unit 3

Let us next consider the connection weights of Hidden Unit 3 (Figure 6-27). As was the case with Hidden Unit 1, Hidden Unit 3 assigns input pitch-classes to equivalence classes related to circles of intervals. For Hidden Unit 3 these equivalence classes involve the three circles of minor thirds.

All four pitch-classes that belong to the first of these circles in Figure 6-15 are assigned the same negative weight (−0.43) in Figure 6-27. All that belong to the second circle of Figure 6-15 have the same weak positive weight (0.07) in Figure 6-27. Finally, all of the pitch-classes that belong to the third circle in Figure 6-15 have the same stronger positive weight (0.33) in Figure 6-27.

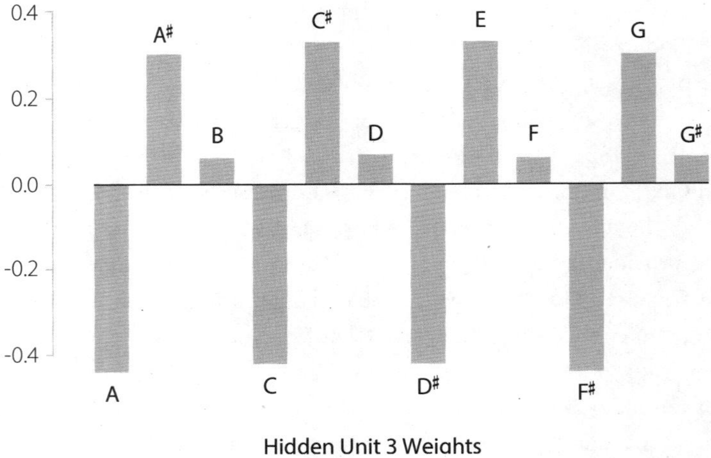

Figure 6-27 The weights of the connections from the input units to Hidden Unit 3.

Unlike Hidden Unit 1, Hidden Unit 3 does not exhibit any obvious balancing between pairs of weights a particular musical interval apart. However, the weights that it assigns to the three different equivalence classes reveal some very interesting properties. If one considers combinations of four different weights, then one discovers specific patterns that cancel net input signals and cause high activity in Hidden Unit 3.

First, it is important to recognize that the value of μ for Hidden Unit 3 is −0.08. This means that for an input pattern to generate a maximum response in this hidden unit, the net input generated by this pattern will be slightly negative.

An examination of different combinations of four weights from Figure 6-27 reveals that there are three different patterns that accomplish this (Figure 6-28).

Figure 6-28 arranges the four different bars representing connection weights in such a way that the balance between negative and positive weights is apparent.

The first combination occurs when a tetrachord contains only one member from the equivalence class assigned a negative weight, only one member from the equivalence class assigned a strong positive weight, and two members from the equivalence class assigned a weak positive weight. This pattern is represented as a stack of four bars on the left of Figure 6-28. When I sum these four weight values, the resulting net input is approximately –0.04, which generates activity of approximately 0.95 in Hidden Unit 3. This pattern appears in four different major seventh chords: Emaj7 [B, D♯, E, G♯], C♯maj7 [C, C♯, F, G♯], Gmaj7 [B, D, F♯, G], and A♯maj7 [A, A♯, D, F]. No other tetrachords, including the other major seventh chords, exhibit this pattern.

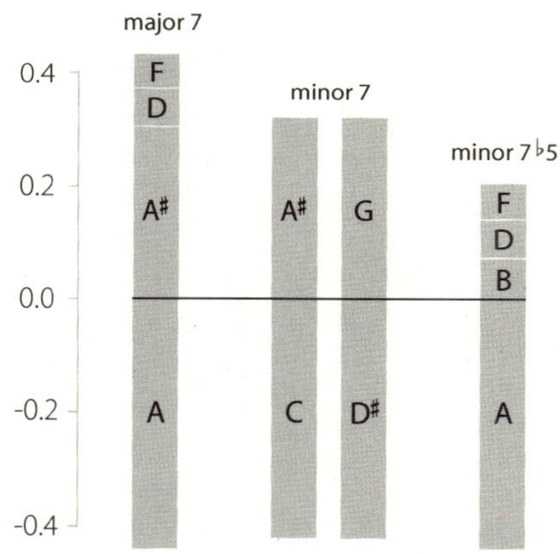

Figure 6-28 Three different combinations of four Hidden Unit 3 weights that produce net inputs close enough to μ to generate high activity.

Interestingly, the other major seventh chords produce moderate activity in this hidden unit (0.63 as shown in Table 6-3). They exhibit a slightly less optimal combination of four weights than the three shown in Figure 6-28. This involves one weak positive weight, one negative weight, and two stronger positive weights or one weak positive weight, one stronger positive weight, and two negative weights. Any of these combinations produces a net input that ranges between –0.47 and 0.30 depending upon which particular weights are included.

The second combination of four Hidden Unit 3 weights that produces high activity involves two pitch-classes that have strong negative weights and two pitch-classes that have strong positive weights. This pattern is illustrated with the group of four bars in the middle of Figure 6-28. This pattern only appears in four different minor seventh tetrachords: F♯min7 [A, C♯, E, F♯], Cmin7 [A♯, C, D♯, G], Amin7 [A, C, E, G], and D♯min7 [A♯, C♯, D♯, F♯].

The third combination of four Hidden Unit 3 weights that produces high activity involves one pitch-class that has a strong negative weight and three pitch-classes that have weak positive weights. This pattern is illustrated with the stack of four bars at the right of Figure 6-28. This pattern only appears in four different minor seventh flat fifth tetrachords: G♯min7♭5 [B, D, F♯, G♯], D min7♭5 [C, D, F, G♯], Fmin7♭5 [B, D♯, F, G♯], and Bmin7♭5 [A, B, D, F].

One type of tetrachord that does not produce high activity in Hidden Unit 3 is the dominant seventh. The highest activity is produced by an input pattern like A♯7: [A♯, D, F, G♯]. Note that this pattern includes three weak positive weights (D, F, G♯), but the fourth weight is a stronger positive weight (A♯). Because most of these weights are weak, this type of input pattern produces a relatively small net input (0.54). However, this net input is extreme enough to reduce Hidden Unit 3 activity to about 0.29. This pattern is true of eight of the 12 different dominant seventh chords. The other four dominant seventh chords include three of the extreme negative weights balanced by only a single weak positive, producing a net input of −1.23 and essentially turning Hidden Unit 3 off.

Finally, while some minor seventh and some minor seventh flat fifth tetrachords cause Hidden Unit 3 to activate, most do not. All of these tetrachords include a pair of pitch-classes that are a minor third apart. As Hidden Unit 3 exploits minor third equivalence, these pitch-classes do not cancel their signals out. Instead, they produce more extreme net input for Hidden Unit 3, reducing its activity. Importantly, the combined effect of a pair of such pitch-classes is not uniform: [A♯, C♯] has a greater effect than does [B, D] because the former pair has more extreme connection weights than the latter pair (see Figure 6-27). This explains why some tetrachords that include at least one minor third can still produce mild activity in Hidden Unit 3 (e.g., 0.27 produced by some minor seventh input patterns).

6.6.4 Hidden Unit 2

Let us finally consider Hidden Unit 2 (Figure 6-29). Although these connection weights have a very regular appearance, they are less musically general than the weights for both Hidden Units 1 and 3. This is because Hidden Unit 2 fulfills a very specialized task for the tetrachord classification network.

Why might one say that the pattern of weights in Figure 6-29 is less musically general than those we have seen earlier in this chapter? One reason is that the weights in Figure 6-29 do not exhibit any systematic assignment of pitch-classes to equivalence classes. For instance, Figure 6-29 begins by suggesting tritone equivalence because the weight for A is nearly identical to the weight for D♯. However, the weights for the next tritone (A♯, E) are not equivalent, nor do they balance. No other systematic equivalences based upon circles of intervals are apparent in this figure either.

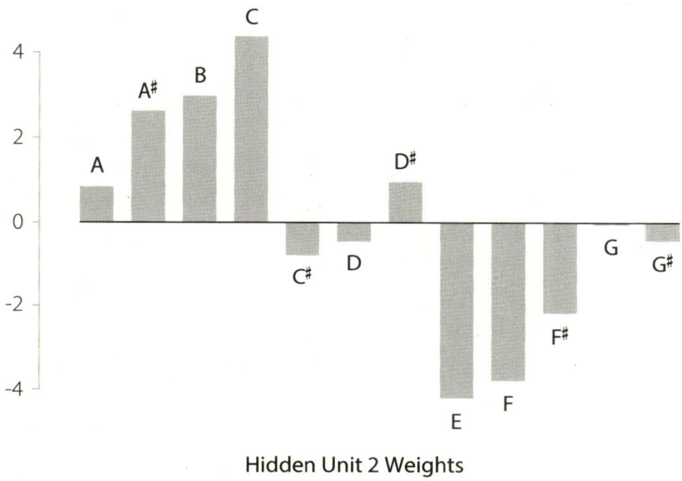

Figure 6-29 The connection weights from the 12 input units to Hidden Unit 2.

We saw earlier that both Hidden Units 1 and 3 organized pitch-classes using interval-based equivalence classes, but also balanced other combinations of pitch-classes related by different intervals. Hidden Unit 2 balances several different pairs of pitch-classes as well. Figure 6-30 illustrates this by presenting the same weights that are in Figure 6-29, but stacking balanced weights on top of each other to highlight their symmetry.

Once again, though, the balancing in Figure 6-30 is not musically systematic. For instance, the balanced pair [A, C♯] is a major third apart, as is the balanced pair [C, E]. However, the balanced pair [A♯, F♯] is a minor sixth apart, while the balanced pair [B, F] is a tritone apart. In short, the connection weights for Hidden Unit 2 seem to balance specific pairs of pitch-classes, and do not balance specific types of musical intervals.

Why does Hidden Unit 2 exhibit properties that are less musically general than those exhibited by the other two hidden units? An answer to this question comes from considering the role of Hidden Unit 2 in arranging input patterns in the hidden unit space.

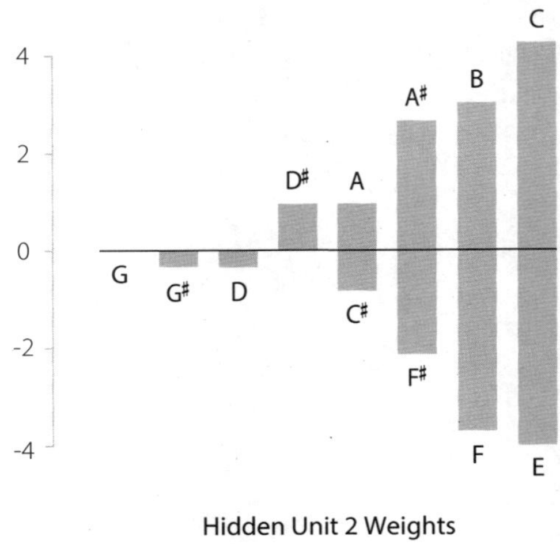

Figure 6-30 The connection weights from the 12 input units to Hidden Unit 2, with balanced weights stacked on top of each other.

To begin, let us consider the hidden unit space in the context of output unit functions. Figure 6-31 attempts to make this context explicit. On its left is a copy of the hidden unit space presented earlier in Figure 6-23. On its right is the same space, but with an additional four planes. These four planes illustrate that in this three-dimensional hidden unit space all of the input patterns that belong to a particular tetrachord type align along a two-dimensional plane that passes through the space.

The planes drawn on the left part of Figure 6-31 are important in terms of how output units function for this particular network. Recall that each output unit is a value unit. This type of unit carves a three-dimensional hidden unit space into decision regions by placing two parallel planes that cut through this space. Any input patterns that fall between these two planes are patterns that turn the output unit on. These planes are very close together, because a value unit is sensitive to a very narrow range of net inputs. The single planes illustrated in Figure 6-31 are important, because they will fall between the two parallel cuts that an output value unit carves through this hidden unit space. This permits the output unit to respond correctly to these patterns by turning on, and by correctly turning off to any other patterns that do not fall between the two cuts.

What is Hidden Unit 2's role in arranging patterns in this space? To answer this question, one can redraw the three-dimensional hidden unit space in Figure

6-31 as a two-dimensional hidden unit space. This two-dimensional space arranges input patterns using Hidden unit 1 and 3 activities as coordinates. In other words, this hidden unit space would exist if Hidden Unit 2 was absent from the multilayer perceptron (Figure 6-32).

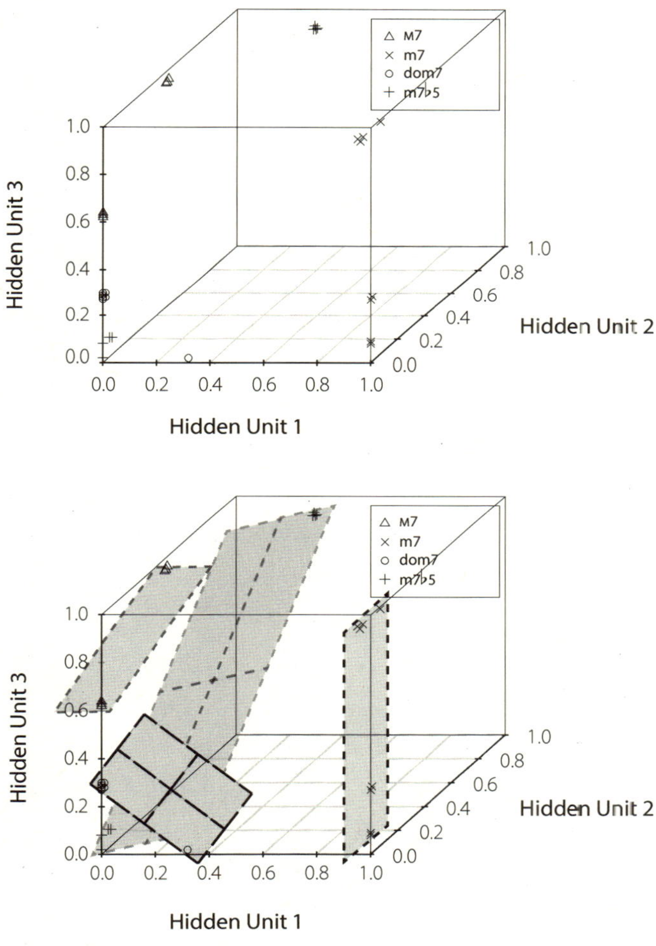

Figure 6-31 The input patterns in their position in the hidden unit space are illustrated on the left.

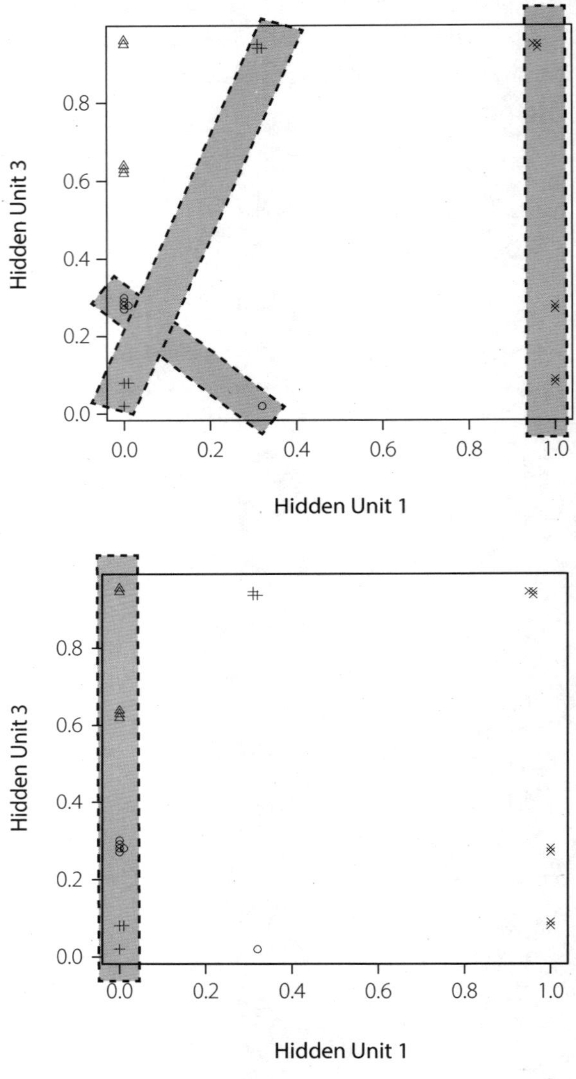

Figure 6-32 A two-dimensional hidden unit space for the input patterns created by removing the Hidden Unit 2 coordinate from Figure 6-31.

When an output value unit confronts a two-dimensional hidden unit space, it does not carve it with parallel planes. Instead, it carves two parallel lines through this space; patterns that fall between the two lines turn the output unit on. The lines are very close together, because an output value unit is sensitive to a very narrow range of net inputs. The hidden unit space on the left side of Figure 6-32 illustrates the parallel cuts that can be made through this space by three of the output units. Each of these three pairs of cuts separates one type of tetrachord from all three of the other types, permitting the output unit to classify the chords. The three cuts illustrated on the left of Figure 6-32 demonstrate that this two-dimensional space arranges input patterns that would permit the network to correctly classify all of the minor seventh, dominant seventh, and minor seventh flat fifth tetrachords.

When an output value unit confronts a two-dimensional hidden unit space, it does not carve it with parallel planes. Instead, it carves two parallel lines through this space; patterns that fall between the two lines turn the output unit on. The lines are very close together, because an output value unit is sensitive to a very narrow range of net inputs. The hidden unit space on the left side of Figure 6-32 illustrates the parallel cuts that can be made through this space by three of the output units. Each of these three pairs of cuts separates one type of tetrachord from all three of the other types, permitting the output unit to classify the chords. The three cuts illustrated on the left of Figure 6-32 demonstrate that this two-dimensional space arranges input patterns that would permit the network to correctly classify all of the minor seventh, dominant seventh, and minor seventh flat fifth tetrachords.

The problem with this two-dimensional hidden unit space, though, is that it does not permit the major seventh tetrachords to be identified. This is illustrated on the right side of Figure 6-32. This graph is the same hidden unit space as the one on the left. In this version of the space, two parallel lines are added to capture the major seventh tetrachords (the triangles). Note that this is the only orientation of these two parallel lines that results in all of the triangles falling between them. However, this positioning of the two lines does not separate the major seventh tetrachords from all of the other types: notice that dominant seventh and minor seventh flat fifth chords also fall between these two lines. This suggests that the functional role of Hidden Unit 2 in the multilayer perceptron is to arrange the major seventh tetrachords in a pattern to separate them from the other chords that they cannot be separated from when Hidden Unit 2 is absent. Looking back at Figure 6-31, this seems to be exactly what the Hidden Unit 2 dimension is adding to the hidden unit space. That dimension appears to capture a handful of major seventh tetrachords and pull them toward the back of the cube. This permits these chords to be arranged along a plane, and permits an output unit to define a decision region that only captures these patterns. The other effect of

Hidden Unit 2 is that it also draws a handful of minor seventh flat fifth tetrachords to the back of the cube. This suggests that these input patterns possess some musical property that is being used by the hidden unit to pull the major seventh tetrachords to the back.

We can confirm this functional account of Hidden Unit 2's role in the network by examining the subset of input patterns that produce higher Hidden Unit 2 activity in the context of the connection weights provided earlier in Figures 6-29 and 6-30. First, Hidden Unit 2 generates moderate to high activity to only eight different tetrachords. This confirms our observation that Hidden Unit 2 only moves a small number of input patterns to permit their correct detection. Second, four of the tetrachords that produce moderate activity in Hidden Unit 2 are major seventh chords whose properties are provided below in Table 6-6. This observation is important because we noted above that the key function of Hidden Unit 2 is to enable this type of chord to be classified; major seventh chords are the only chords that cannot be correctly separated in the two-dimensional pattern space.

Table 6-6 The four major seventh tetrachords and then the four minor seventh flat five tetrachords that produce moderate activity in Hidden Unit 2.

Chord	Notes				Weights				Net	H2	Similar
A#maj7	A	A#	D	F	0.92	2.69	−0.39	−3.77	−0.55	0.49	**Bm7flat5**
Gmaj7	B	D	F#	G	3.03	−0.39	−2.19	−0.05	0.40	0.49	**G#m7flat5**
C#maj7	C	C#	F	G#	4.36	−0.75	−3.77	−0.40	−0.55	0.48	**Dm7flat5**
Emaj7	B	D#	E	G#	3.03	0.95	−4.15	−0.40	−0.56	0.47	**Fm7flat5**
Bm7flat5	A	B	D	F	0.92	3.03	−0.39	−3.77	−0.18	0.96	**A#maj7**
G#m7flat5	B	D	F#	G#	3.03	−0.39	−2.19	−0.40	−0.19	0.96	**Gmaj7**
Dm7flat5	C	D	F	G#	4.36	−0.39	−3.77	−0.40	0.06	0.95	**C#maj7**
Fm7flat5	B	D#	F	G#	3.03	0.95	−3.77	−0.40	−0.20	0.95	**Emaj7**

Note. The grey cells indicate the one difference between the notes of each of these major seventh tetrachords and the minor seventh flat five tetrachord to which it is similar. The "Net" column provides the net input produced for Hidden Unit 2, and the "H2" column provides Hidden Unit 2 activity.

Third, the four tetrachords that produce high activity in Hidden Unit 2 are all minor seventh flat fifth chords that are nearly identical to the four major seventh chords that produce moderate activity in this same unit. They are nearly identical in several respects. They share three pitch-classes with one of the major seventh tetrachords. Their remaining pitch-class is only a minor second away from the fourth pitch-class. Finally, the connection weight associated with the fourth pitch-class has a similar value to the connection weight associated with the fourth (i.e., the dissimilar) pitch-class in the major seventh chord. Table 6-6 provides the properties of the four minor seventh flat fifth tetrachords. The grey cells in Tables 6-6 indicate the single difference, in either a pitch-class or a connection weight, between a major seventh tetrachord and its similar minor seventh flat fifth tetrachord.

The high specificity of the connection weights presented in Figures 6-29 and 6-30 now make perfect sense in light of the properties detailed in the two tables above. First, these connection weights capture very specific relationships (and not general musical properties) because all Hidden Unit 2 really has to do is move four major seventh tetrachords away from the others in the three-dimensional hidden unit space. If this unit detects properties that are more general, then this would affect the position of a larger number of tetrachords. Second, the specific balancing observed in Figure 6-30 nicely accomplishes the main task of Hidden Unit 2. The A♯maj7 chord produces moderate activity in this unit because the weights from A and from D roughly balance one another, as do the weights from A♯ and F. A similar rough balancing of pairs of pitch-classes is true for Emaj7. The remaining two major seventh chords balance one extreme positive weight (C for C♯maj7, B for Gmaj7) with a combination of three negative weights. Third, the four minor seventh flat fifth tetrachords that produce high activity in Hidden Unit 2 do so because they are each nearly identical to one of the major seventh chords that this unit moves. They differ in only one pitch-class, and this difference is only a semitone. This in turn means that one structural property of the weights in Figure 6-29 is that pairs of pitch-classes a minor second apart ([A♯, B], [C♯, D], [E, F] and [G, G♯]) must have similar weight values. An inspection of this figure indicates that this property is indeed apparent.

6.7 Summary and Implications

This chapter began by describing a multilayer perceptron for classifying triads regardless of their key or inversion. This network revealed interesting patterns of connection weights between its input and hidden units. In particular, its connection weights organize pitches and pitch-classes into different subsets. For instance,

Hidden Unit 1 assigns weights so that [A, C, D♯, F♯] define one subset, and similarly uses its weights to organize [A♯, C♯, E, G], [B, D, F, G♯], and [C, D♯, F♯, A] as three other subsets. That is, pitch-classes that belong to the same subset have a very similar weight, but pitch-classes that belong to different subsets have very different weights. We see similar sorts of organizations, involving different subsets of pitch-classes, in the other hidden units as well.

The chapter then proceeded to describe the various interval cycles created by moving set distances from pitch to pitch along a piano keyboard. These interval cycles are a standard element of post-tonal music theory (Roig-Francolí, 2008). I noted that they also provide a convenient formalism for the subsets of pitches and pitch-classes picked out by connection weights. This in turn led to the notion of strange circles. A strange circle is a set of pitch-classes that belong to the same interval cycle. The circle is strange because each member of the same strange circle has the same connection weight. This means that a strange circle is an equivalence class of pitch-classes. It captures a set of pitch-classes that are distinct in Western music, but identical from the perspective of a hidden unit in a network.

The chapter ended by providing a detailed interpretation of the internal structure of a multilayer perceptron that learned to classify tetrachords into four different types. This network provided another example of using strange circles. Both Hidden Units 1 and 3 of this network organize pitch-classes into equivalence classes based upon interval cycles. Hidden Unit 1 assigns the same weight values to two pitch-classes that belong to the same circle of tritones. Hidden Unit 3 assigns the same weight values to three pitch-classes that belong to the same circle of minor thirds. Furthermore, the particular weight values that both Hidden Units 1 and 3 assign to each equivalence class are very systematic. Their values permit pitch-classes separated by other musical intervals to balance, increasing hidden unit activation.

Taken together, these two above observations indicate that Hidden Units 1 and 3 detect general musical properties that permit varied and useful responses to pitch-class combinations that are either part of, or not part of, specific tetrachord structures. Indeed, Figure 6-32 indicates that these two hidden units alone are capable of supporting the correct identification of all of the members of three of the four types of tetrachords. The network's remaining hidden unit, Hidden Unit 2, detects very specific properties (i.e., properties related to a small number of individual chords, and not to a larger set or chord types) that serve to arrange the major seventh chords in hidden unit space in such a way that they can be correctly identified. The specific properties detected by this hidden unit also capture four different minor seventh flat fifth chords. There is a strong musical relationship between these chords and the major seventh chords segregated by Hidden Unit 2.

To relate this interpretation to material covered in Chapter 4, these three hidden units provide another example of coarse coding. None of the hidden units detects a specific property that is consistent with only one type of tetrachord: two or more different types of tetrachords can produce high activity in any of the hidden units. However, when we consider the activities produced by an input pattern in the hidden units simultaneously, we can identify the input pattern's type.

The connection weights for the two more general hidden units (Hidden Units 1 and 3) of the network interpreted in Section 6.6 have one interesting implication to keep in mind for later network interpretations: both of these hidden units use their weights to organize pitch-classes by more than one musical interval. The reason that this is possible is because some of the interval cycles introduced in Section 6.3 relate hierarchically to others. For example, each of the circles of major seconds contains the pitch-classes that belong to two of the circles of major thirds. Similarly, each of the three circles of major thirds contains the notes that belong to two of the circles of tritones. These hierarchical relationships permit one set of connection weights to organize input pitch-classes in complex ways. For example, one hidden unit could use the sign of the connection weight to separate pitch-classes into the two circles of major seconds. However, variations in magnitudes of these same weights can simultaneously be used to organize the pitch-classes into circles of tritones because of a hierarchical relationship between the two. In Chapter 7, I explore a more complex network trained on an elaboration of the tetrachord task, and discover that many of its hidden units exploit the hierarchical relationships among strange circles in their representation of input pitch-classes.

7

Classifying Extended Tetrachords

7.1 Extended Tetrachords

7.1.1 Extended Chords

Chapter 6 described a multilayer perceptron for classifying four different types of tetrachords, and detailed its internal structure. In this chapter, I turn to a more complicated musical problem, one that involves a larger set of different types of tetrachords. Because this problem is more complex, the multilayer perceptron that solves it requires more hidden units. However, these hidden units also organize inputs into a variety of strange circles that assist in interpreting the network's internal structure. The four tetrachords explored in Chapter 6 were all examples of added note tetrachords. That is, each tetrachord started as a triad built from three different notes that belonged to a musical scale. I created a tetrachord by adding a fourth note, which also belonged to the scale, to the triad.

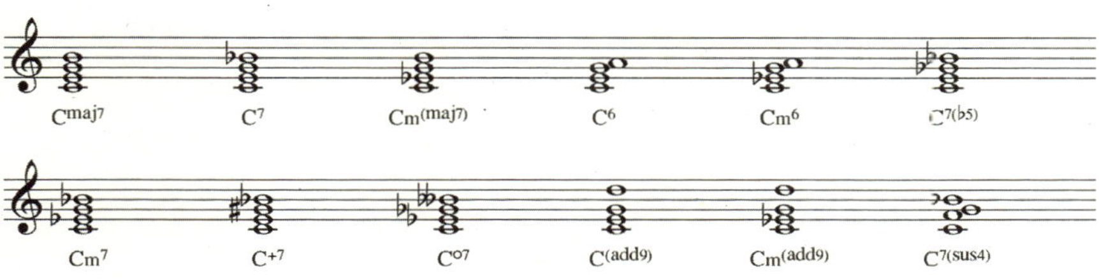

Figure 7-1 Musical notation for 12 different types of tetrachords, each using C as the root note.

A different approach to building tetrachords produces a greater variety of chord types. One begins with a triad formula. For instance, if one takes the first, third, and fifth notes of the C major scale (C, E, G), the result is the C major triad. Therefore, the formula for the C major triad is 1-3-5. Adding the seventh note of the scale, B, produces the C major seventh tetrachord, which follows the formula 1-3-5-7.

More chords can be created by manipulating formulae similar to the one provided in the previous paragraph. For instance, one could flatten the third and the fifth note in the formula 1-3-5-7. This produces the formula 1-♭3-5-♭7; if C is the root then this formula produces the set of notes [C, E♭, G, B♭], which defines the C minor seventh tetrachord. Note that the flattened third and seventh notes do not belong to the C major scale.

In jazz, one often finds extended chords that use formulae that add notes that fall beyond the octave range of a major scale. For example, if one adds the D that is an octave higher than the second note in the C major scale to the C major triad, then one produces the Cadd9 tetrachord [C, E, G, D]. The formula for this chord is 1-3-5-9.

Figure 7-1 provides the musical notation, and the musical chord symbol, for 12 different types of tetrachords. Each of these example tetrachords uses C as the root note of the chord. We saw four of these tetrachord types earlier in Chapter 6. The other eight are new; Table 7-1 provides the formula for each.

Table 7-1. The names and formulas for twelve different types of tetrachords.

Tetrachord type	Formula	Example and notation	Forte number
Major seventh	1-3-5-7	Cmaj7	4-20(12)
Dominant seventh	1-3-5-♭7	C7	4-27
Minor, major seventh	1-♭3-5-7	Cm(maj7)	4-19
Sixth	1-3-5-6	C6	4-26(12)
Minor sixth	1-♭3-5-6	Cm6	4-27
Seventh, flat five	1-3-♭5-♭7	C7flat5	4-25(6)
Minor seventh	1-♭3-5-♭7	Cm7	4-26(12)
Augmented seventh	1-3-♭5-♭7	Caug7	4-24(12)
Diminished seventh	1-♭3-♭5-♭♭7	Cdim7	4-28(3)
Added ninth	1-3-5-9	Cadd9	4-22
Minor added ninth	1-♭3-5-9	Cm(add9)	4-14

Tetrachord type	Formula	Example and notation	Forte number
Seventh, suspended fourth	1-4-5-♭7	C7sus4	4-23(12)

Note. An example of each chord is provided in Figure 7-1. The final two columns provide the notation of an example chord that belongs to the type, as well as the classification number for the chord type from Forte's (1973) set theory.

The formulae provided in Table 7-1 work in the context of any major scale. The numbers in each formula refer to a note's position in a particular scale. That is, 1 is the first note in a particular scale, 3 is the third note in a particular scale, and so on. This means that there are 12 different versions of each of the chord types listed in Table 7-1: one for each of the 12 possible major scales.

When I use these formulae to create tetrachords in different keys, some interesting relationships between chords arise. Consider the 6 chord whose formula is 1-3-5-6. In the context of the C major scale this produces the C6 chord whose notes are [C, E, G, A]. Now consider applying the formula for the minor seventh tetrachord (1-♭3-5-♭7) in the context of the A major scale. This produces the Am7 chord whose notes are [A, C, E, G]. Note that these notes are identical to those of C6; musically speaking, Am7 is identical to an inversion of C6. Similarly, the dominant seventh chord is the inversion of a minor sixth tetrachord in a different key.

In other words, the same set of four pitch-classes can have more than one chord name. If I train a network to identify tetrachord types, then it must generate both of these chord names to one set of four input pitch-classes. Table 7-1 also provides the Forte numbers of each of these chord types. Forte numbers are a system for classifying different musical entities that is derived from using mathematical set theory (Forte, 1973). Note that different tetrachord names for the same set of input pitch-classes have the same Forte number in Table 7-1, indicating that the chords have the same basic structure in spite of the fact that they have different names.

When I train a multilayer perceptron to classify the 12 different types of tetrachords in Table 7-1, I will again use pitch-class representation. Because of this, notes in extended chords like the added ninth chord are moved back into the range of a single octave. It is therefore useful to represent the various tetrachords in a visual format. One can illustrate a tetrachord in a circle of minor seconds by drawing in four spokes that represent the four pitch-classes present in a particular chord. Drawing such a diagram will illustrate a particular chord in the context of a specific major key. However, this diagram represents the structure of a tetrachord type for any key:

if one rigidly rotates the spokes to a different position in the circle, then it will provide the notes for the same type of tetrachord, but relative to some other musical key. Figure 7-2 provides pitch-class diagrams for the 12 tetrachords from Figure 7-1.

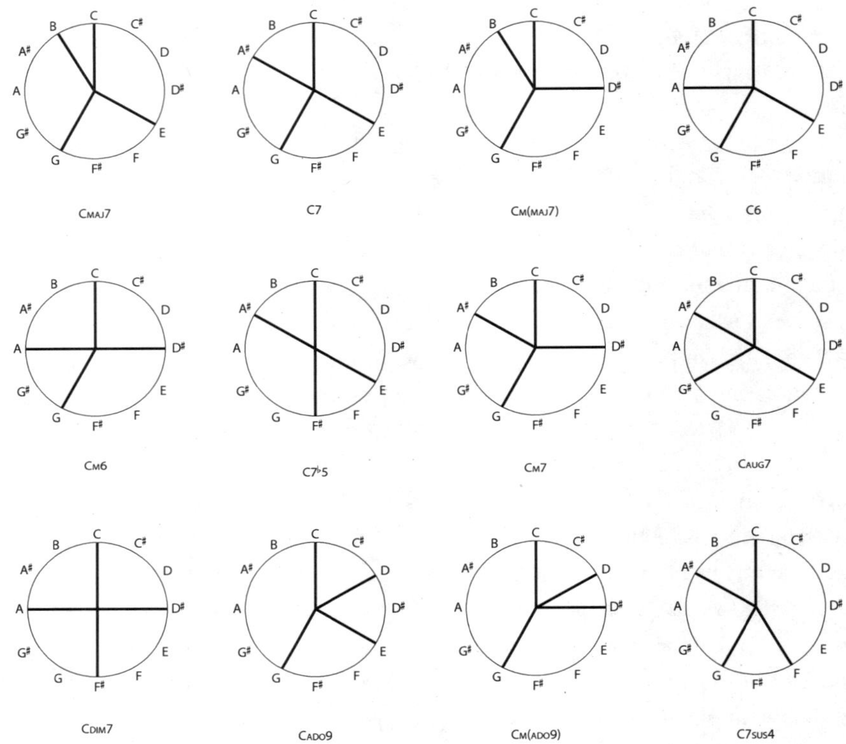

Figure 7-2 Pitch-class diagrams of the 12 tetrachords from the score in Figure 7-1.

All of the chords presented in Figure 7-2 are defined with respect to the C major scale. The structure of the spokes in the diagrams provides an interesting perspective on the similarities and differences between various tetrachord types. For instance, it is immediately apparent that both the diminished tetrachord and the seventh flattened fifth tetrachord include two pairs of notes that belong to the same circle of tritones, because both diagrams include two long spokes that bisect the circle. Similarly, one can see the similarity in spoke structure between the minor seventh and the sixth tetrachords, as well as between the seventh and the minor sixth tetrachords. In the next section, we will describe training a multilayer perceptron to identify these 12 different types of tetrachords.

7.2 Classifying Extended Tetrachords

7.2.1 Task

Our goal is to train an artificial neural network, when presented with four notes that define a tetrachord, to identify the type of tetrachord, ignoring the tetrachord's key. The difference between the current network and those described in Chapter 6 is that the current network learns to classify input chords into 12 different categories instead of only four. After being trained on this task the multilayer perceptron typically turns one output unit "on" to identify tetrachord type, and turns the remaining 11 output units "off," when presented a tetrachord. The exception to this occurs when two different tetrachord types (e.g., 6 and m7) apply to the same four input pitch-classes. In this situation, the network turns on both of the appropriate output units, and turns the remaining ten output units off.

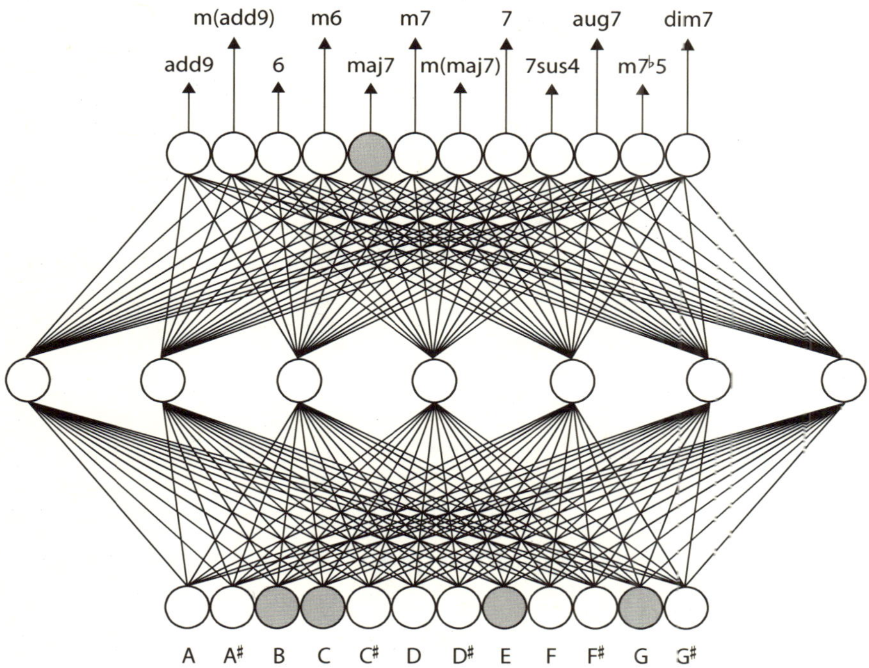

Figure 7-3 The architecture of the multilayer perceptron trained to identify 12 different types of tetrachords.

7.2.2 Network Architecture

Figure 7-3 illustrates the architecture of the current network. It uses 12 input units to represent input pitch-classes. It requires 12 output units to identify all of the tetrachord types from Figure 7-1. The network requires seven hidden units to find a solution to the extended tetrachord problem. All of the output units and all of the hidden units in the network are value units.

7.2.3 Training Set

The training set consists of 144 stimuli: the 12 different tetrachords that can be created in the context of a particular major scale (see Figure 7-1). I create these tetrachords for each of the 12 different major scales. Each is encoded as an input pattern in which four input units are activated with a value of one, and the remaining eight input units are all activated with a value of zero. I pair each input pattern with an output pattern that indicates the tetrachord type to which the input pattern belongs. I train the network to turn on the output units that represent the input patterns type(s), and to turn all other output units off.

7.2.4 Training

The multilayer perceptron in Figure 7-3 is trained with the generalized delta rule developed for networks of value units (Dawson & Schopflocher, 1992) using the Rumelhart software program (Dawson, 2005). During a single epoch of training each pattern is presented to the network once; the order of pattern presentation is randomized before each epoch.

All connection weights in the network are set to random values between −0.1 and 0.1 before training begins. In the network described in detail below, each μ is initialized to zero but is then modified by training. A learning rate of 0.01 is employed. Training proceeds until the network generates a "hit" for every output unit for each of the 144 patterns in the training set. Again, a "hit" is defined as activity of 0.9 or higher when the desired response is one or as activity of 0.1 or lower when the desired response is zero. A network that contains seven hidden value units solves this problem readily, typically converging after between 7000 and 10,000 epochs of training. The example network described in more detail in the next section converges after 7236 epochs of training.

7.3 Interpreting the Extended Tetrachord Network

This section provides an analysis of the connection weight structure of each of the hidden units in the trained network. This analysis reveals a number of interesting musical regularities in this network's structure. However, it is very detailed. The reader, who is less interested in these details, and more interested in a general summary of these results, will find this summary in Section 7.4.

7.3.1 Jittered Density Plots

The extended tetrachord network is the most complicated one that we have encountered in this book. This is because it has seven hidden units, making it very difficult to orient an interpretation by graphing the hidden unit space. For this reason, I will begin to interpret the network by examining two different characteristics of each hidden unit: the weights of the connections that feed into a hidden unit and the activity produced by the hidden unit when it is presented each of the 144 input patterns.

With respect to patterns of connectivity, each of the hidden units organizes input pitch-classes into some of the strange circles from Chapter 6. This is particularly helpful for interpreting this more complicated network. Instead of considering the effect of the 12 different pitch-classes on the hidden unit, we can consider smaller sets of pitch-classes that are treated as being equivalent. For example, we will see that an account of Hidden Unit 1's role in the network can be achieved by considering input pitch-classes as belonging to one of the two circles of major seconds, or as belonging to one of the six circles of tritones.

With respect to hidden unit activity, I will take advantage of a characteristic that is frequently exhibited by value units (Berkeley et al., 1995), although in some cases it may be found in other types of processors (Berkeley & Gunay, 2004). When the activities of a hidden value unit are graphed using a jittered density plot, this plot is often organized into different bands. Each band contains a subset of input patterns that share certain properties which, when identified, help in understanding the features being detected by the hidden unit. Let us describe the general use of banded jittered density plots in more detail before using them to interpret the extended tetrachord network.

A jittered density plot can be thought of as a one-dimensional scatter plot. Consider producing a jittered density plot for the activities generated by one hidden unit to each of the patterns of a training set. Each pattern is represented by one dot in the plot. The position of the dot along the x-axis of the graph provides the activity produced in the hidden unit by that pattern. The position of the dot along the y-axis is a random number that has no meaning; this random jittering prevents different dots in the plot from overlapping as much as possible.

An example jittered density plot for Hidden Unit 1 of the current network is provided in Figure 7-4 below. Note that the x-axis ranges from zero to one, because this is the activity range of a value unit. There are 144 different dots in this plot, one for each of the 144 tetrachords in the training set.

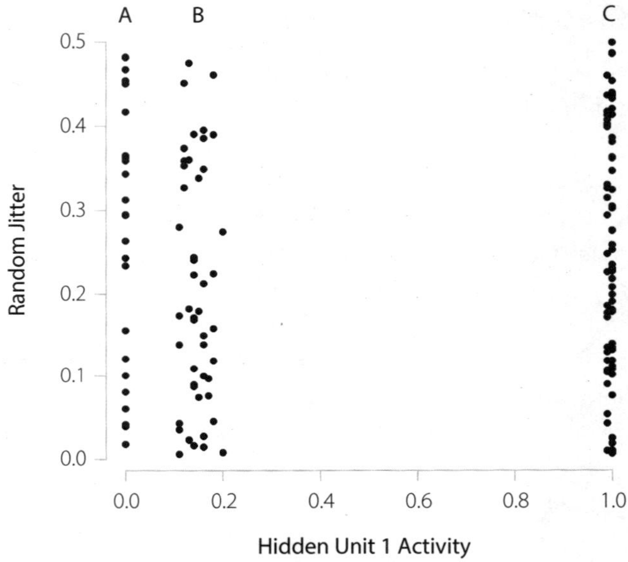

Figure 7-4 The jittered density plot for Hidden Unit 1 in the extended tetrachord network.

Berkeley et al. (1995) discovered that in many cases the jittered density plots of hidden value unit activities organize themselves into distinct bands. This is true of the jittered density plot in Figure 7-4. It is organized into three different bands: in Band A, 24 of the input patterns generate zero activity in this unit; in Band B, 48 of the patterns generate activity that ranges between 0.11 and 0.20, and in Band C, the remaining 72 patterns generate activity between 0.99 and 1.

Berkeley et al. (1995) discovered that patterns that belong to the same band in a jittered density plot share certain properties. By examining the characteristics of just the subset of patterns that fall into one band, one can interpret the features they share and use these features to determine the unit's function in the network (Dawson et al., 1997; Dawson et al., 2000b; Dawson & Piercey, 2001). Figure 7-4 demonstrates that distinct banding is present when the activities of one of the extended tetrachord's hidden units are graphed in a jittered density plot. Fortunately for us, banding is present for almost all of the hidden units of this network. I will take advantage of this banding by taking just those input patterns that fall into a particular band and determining what features these tetrachords have in common. Furthermore, this interpretation is informed by our understanding of the strange circles found in the connection weights in each hidden unit. Together these two properties will lead to a detailed understanding of the internal structure of the extended tetrachord network. As was the case earlier in Section 6.6, I will discuss the hidden units out of order so that the units that are easier to interpret are described before those that are more complicated.

7.3.2 Hidden Unit 1

Figure 7-5 provides a graph of the connection weights that feed into Hidden Unit 1 from the 12 input pitch-class units, combined with the jittered density plot from Figure 7-4. It is obvious from Figure 7-5 that this hidden unit organizes input signals in terms of strange circles.

First, all of the positive weights come from pitch-classes that belong to one of the circles of major seconds, and all of the negative weights come from pitch-classes that belong to the other circle of major seconds. Second, if one examines the set of six positive weights, then it becomes apparent that there is some variation in strength. This variation occurs because this hidden unit assigns the identical weight to pairs of pitch-classes that belong to the same circle of tritones. This variation in weights permits the hidden unit to distinguish one circle of tritones from another. This is also true of the six negative weights.

What does this hidden unit detect? To begin, let us note that at the end of training this unit's μ has a value of −0.01, indicating that it turns on when it receives a near-zero net input. With this fact in mind, and recognizing that Hidden Unit 1 appears to use equivalence classes involving circles of minor seconds and circles of tritones, let us consider the patterns that fall into each of the three bands of the jittered density plot.

First, consider the subset of patterns that belong to Band A in Figure 7-5. There are only two types of tetrachords in this subset: all of the aug7 chords and all of the 7♭5 chords. What do these tetrachords have in common? Each chord includes four pitch-classes that all belong to only one of the circles of major seconds. As a result, all four of the signals sent to Hidden Unit 1 by one of these chords pass through weights that all have the same sign. These signals cannot cancel one another out;

the hidden unit will receive either an extreme positive or an extreme negative net input which causes it to turn off because of its near zero μ.

Now consider the subset of patterns that fall into Band C in Figure 7-5. These patterns consist of all the 6, 7sus4, dim7, m (add9), m7, and maj7 tetrachords. What does this large collection of different types of chords have in common?

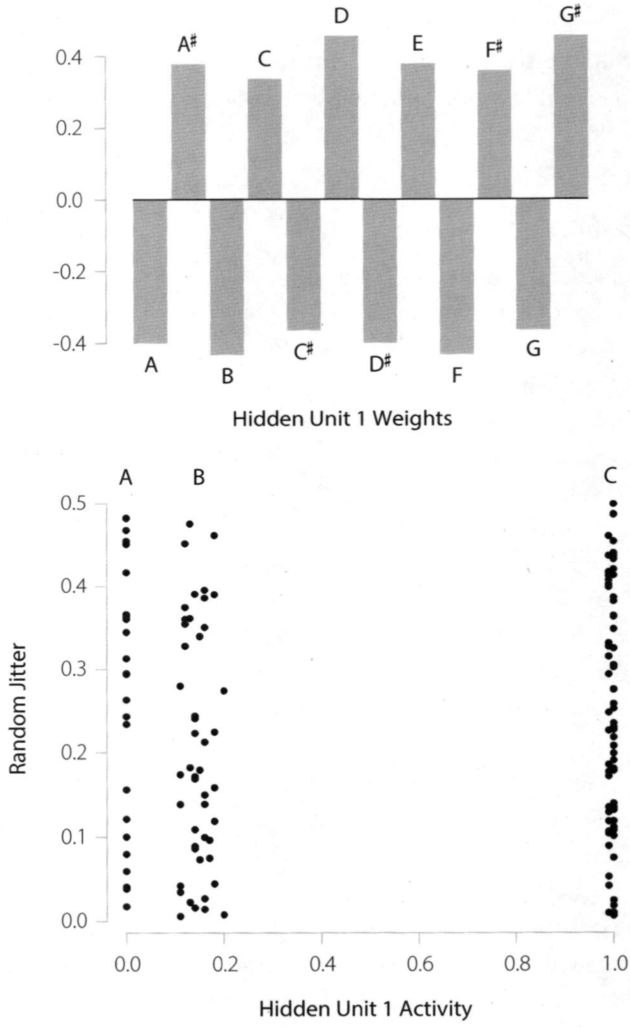

Figure 7-5 The connection weights and the jittered density plot for Hidden Unit 1.

First, all of these tetrachords have two pitch-classes that belong to one circle of major seconds and two others that belong to the other circle of major seconds. This permits the signals sent from these chords to cancel each other out, producing a near-zero net input, and turning Hidden Unit 1 on. Second, the tetrachords that belong to this band (with the exception of the °7 chords, which are a special case) include pitch-classes that each belong to a different circle of tritones. As a result, four different circles of tritones are represented in each chord. For any chord, which four circles of tritones are represented is important: two of the sampled circles have negative weights, while the other two have positive weights.

As a result, one finds in these tetrachords two specific patterns of tritone sampling. These are illustrated in Figure 7-6A and Figure 7-6B. In these figures, each tritone circle is a line that bisects the pitch-class diagram; there are six in each figure. Tritone circles that have a pitch-class that belongs to the pitch-classes of these tetrachords are represented as solid lines; dashed lines indicate tritone circles that are not represented. In the first pattern exhibited by the chords that belong to Band C (Figure 7-6A), the tetrachords contain pitch-classes from four adjacent tritone circles. Note that because weights in the network are organized by circles of major seconds, two negative and two positive weights are involved in these chords, producing zero net input. The same is true for the second pattern (Figure7-7B): a tetrachord contains pitch-classes from two adjacent tritone circles, not from the next, and then contains pitch-classes from the next two adjacent tritone circles. Only the diminished seventh (°7) tetrachords fail to exhibit this pattern, but this is because they represent a special case of Figure 7-6B: they sample two circles of tritones twice, and these two samples are from circles that are 90° apart in the diagram (see Figure 7-2).

The importance of which circles of tritones are represented by a tetrachord's pitch-classes emerges when we consider the final band of patterns that produce weak activity in Hidden Unit 1 (Band B, Figure 7-5). This band includes all of the remaining types of tetrachords (7, add9, m (maj7), m6). Half of these chords fall into this band because they represent three different circles of tritones, not four. In other words, they contain one pitch-class each from two different circles of tritones, and contain two pitch-classes from a third. As a result, the input signals do not cancel one another out.

However, the remaining tetrachords that belong to this band sample pitch-classes from four different tritone circles. Why do these chords not turn Hidden Unit 1 on? The answer to this question is that they represent these tritone circles following a different pattern than the two discussed above. As shown in Figure 7-6C, they contain pitch-classes from three adjacent tritone circles, skip the next, and then contain a pitch-class from the next. This pattern of sampling produces an unbalanced signal, generating weak activity in this hidden unit.

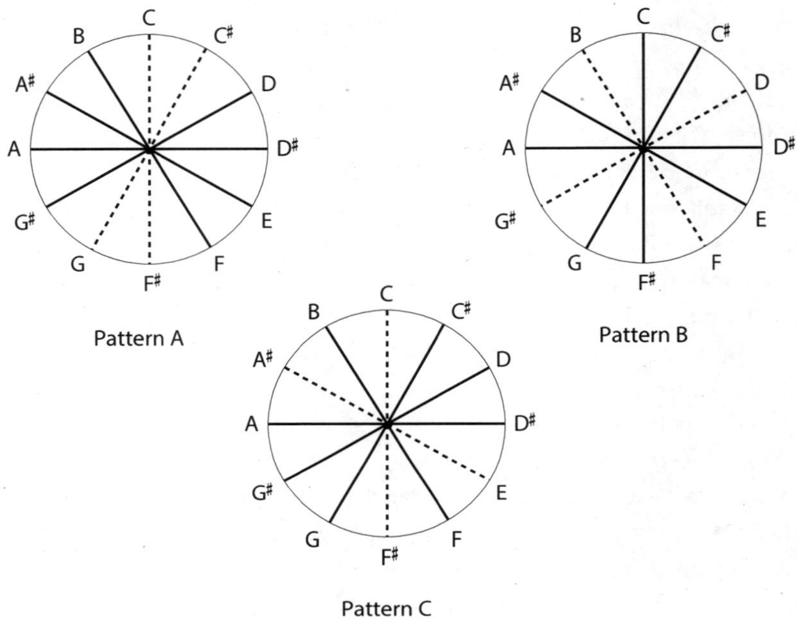

Figure 7-6 Three patterns of tritone sampling for tetrachords. A and B are patterns that turn Hidden Unit 1 on; C is a pattern that generates weak activity in Hidden Unit 1.

7.3.3 Hidden Unit 2

Figure 7-7 provides the connection weights and the jittered density plot for Hidden Unit 2 of the extended tetrachord network. This hidden unit organizes input pitch-classes into circles of minor thirds, assigning a weight of 0.79 to those pitch-classes that belong to the first circle, a weight of −0.07 to those pitch-classes that belong to the second, and a weight of −0.50 to those pitch-classes that belong to the third. At the end of training, the value of μ for this unit is −0.13.

This jittered density plot is similar to the one for Hidden Unit 1, as it is organized into three distinct bands. The first is near zero, the second is between 0.2 and 0.4, and the third is between 0.8 and 1.0. The bands for Hidden Unit 2 are slightly more dispersed than those observed for Hidden Unit 1.

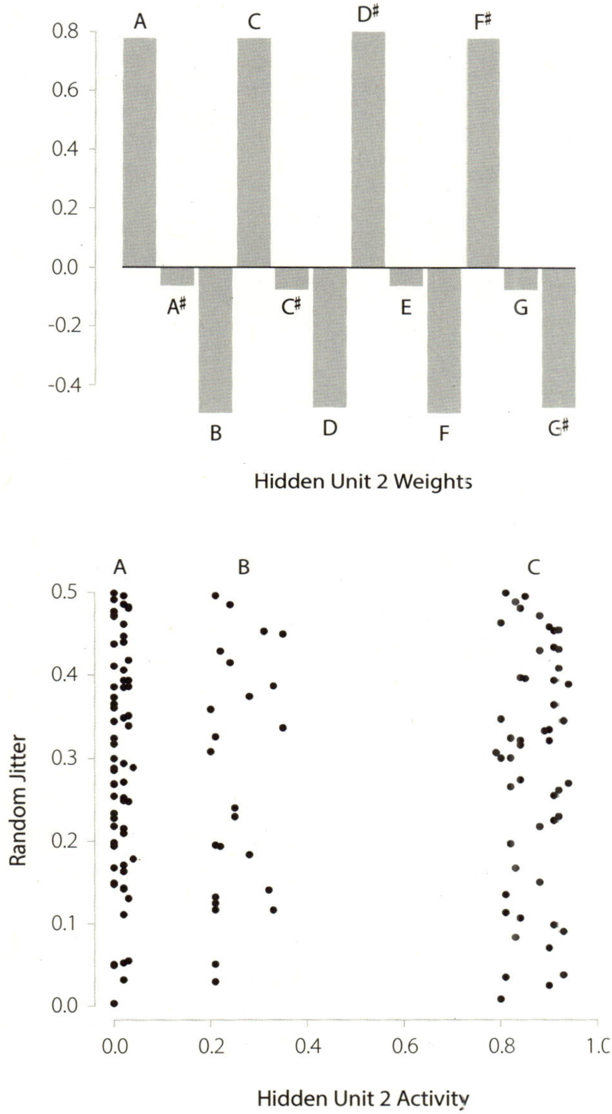

Figure 7-7 The connection weights and the jittered density plot for Hidden Unit 2.

Let us first consider the patterns that belong to Band C in Figure 7-7. There are 52 such patterns, representing 7sus4, add9, aug7, dim7, m (add9), m (maj7), and maj7 tetrachords. Interestingly, the band does not capture all instances of each chord type: it captures four instances of the diminished seventh chord and eight instances of each of the other chord types. Whatever property belongs to the chords in this band does not characterize all 12 instances of each chord type.

What properties do the tetrachords that belong to this band share? All of these tetrachords (except the diminished sevenths, which are a special case) select pitch-classes from each of the three circles of minor thirds. That is, they select one pitch-class from each of two of these circles, and select two pitch-classes from the third circle. Furthermore, 24 of the tetrachords in Band C include one pitch-class associated with a weight of 0.79, a second associated with a weight of −0.07, and two pitch-classes associated with a weight of −0.50. This results in a net input of about −0.30, which is close enough to μ to produce activity of about 0.90. Another 24 of the tetrachords include two pitch-classes associated with a weight of −0.07, and two others associated with each of the other two weights. This produces a net input of 0.14, resulting in activity of just over 0.80.

The diminished seventh chords that fall in this band are a special case, because they are composed of all four pitch-classes that are associated with a weight of −0.07, which all belong to the same circle of minor thirds. These four weights sum to −0.28, a net input that produces activity of 0.88 in Hidden Unit 2.

Why do we only find subsets of different tetrachord types in this band? The structure of the four diminished seventh tetrachords provides an answer to this question. The other eight diminished seventh chords are composed of four pitch-classes that all belong to one of the other two circles of minor thirds. When I sum these weights, the resulting net input is too extreme to produce high activity in Hidden Unit 2. This removes them from this band.

A similar story holds for the other types of tetrachords in this band. Recall that the band captures eight instances of each type, but four other instances do not belong to the band. This is because the specific set of weights for Hidden Unit 2 is such that these subsets of tetrachords generate an extreme net input that removes them from the band. For example, Gmaj7, Cmaj7, Fmaj7, and G♯maj7 are similar to all of the other major seventh chords in that they include two pitch-classes from one circle of minor thirds and one from each of the other two. However, given the weights for Hidden Unit 2, their particular combination of notes produces a net input that removes them from the band.

In particular, each of these chords includes two pitch-classes from the circle of minor thirds assigned a weight of 0.79 by this unit, and one pitch-class from each of the other two circles. As a result, these four major seventh chords generate a net input of one, which turns Hidden Unit 2 off. This separates these four tetrachords from the other eight that fall in the high band. A similar account holds for all of the other chords that belong to a tetrachord type captured by the band, but which are not part of the band.

Let us next consider the band of patterns that produce weak activity (ranging between 0.2 and 0.4) in Hidden Unit 2 (Band B). There are 24 such patterns, representing m6, 6, m7, 7, and 7♭5 tetrachords. Again, the band does not capture all instances of each chord type. All of the chords that fall in this band share one property: they do not include a pitch-class from one of the three circles of minor thirds. Either they include three pitch-classes from one circle and a fourth from one other, or they include two pitch-classes from one circle and two others from another. In either case, the weights associated with these sets of pitch-classes cannot cancel each other out; these chords produce net inputs of either −0.72 or 0.58.

From the discussion above, it appears that high activity in Hidden Unit 2 indicates that it detects a tetrachord characterized by one of two different patterns. One pattern involves four pitch-classes associated with a particular combination of connection weights (one strong positive, one weak negative, two strong negatives). The second pattern involves four pitch-classes each of which is associated with a weak negative connection weight.

The patterns that belong to Band A in Figure 7-7 produce zero activity in Hidden Unit 2 because they fail to exhibit either of these combinations of weights. As a result, the 68 patterns that belong to this band represent all 12 different types of tetrachords in the training set.

When banding in the jittered density plots of value units was first discovered (Berkeley et al., 1995), it was noted that patterns associated with a band associated with near-zero activity were patterns that did not share any defining positive feature. Instead, they shared a negative feature: they all lacked the features that the hidden unit detects, and which produce higher activity. As a result, in many cases a detailed interpretation of the features of patterns that belong to a "zero band" is neither informative nor possible. Band A in Figure 7-7 is an example of this situation.

7.3.4 Hidden Unit 4

Band C for the jittered density plot of Hidden Unit 2 (Figure 7-7) indicates that this unit generates high activity to a number of different types of tetrachords. However, for each of these different types, it generates this high activity to only eight of the 12 possible instances. What does the network do to the four instances of each chord type omitted from this band in Hidden Unit 2? They are the only chords that produce high activity in Hidden Unit 4!

Figure 7-8 provides the connection weights and the jittered density plot for Hidden Unit 4. Examining the weights indicates that this hidden unit, like Hidden Unit 2, organizes input pitch-classes into circles of minor thirds, assigning a weak negative weight to those pitch-classes that belong to the first circle, a more negative weight to those pitch-classes that belong to the second, and a strong positive weight to those pitch-classes that belong to the third. At the end of training, the value of μ for this unit is −0.06. The weights also indicate that pitch-classes are also organized into equivalence classes based upon circles of tritones: pitch-classes that are in the same circle of tritones are assigned identical weights. Indeed, this organization is cleaner than the organization in terms of circles of minor thirds, because there is some variation of weight values assigned to pitch-classes in the same circle of minor thirds.

The lower part of Figure 7-8 indicates that the jittered density plot of Hidden Unit 4 is organized into two fairly broad bands: patterns that belong to Band A generate activity that ranges between 0.00 and 0.50, while patterns that belong to Band B generate activity that ranges between 0.80 and 1.00. I consider these as different bands because there is a large space in the graph between them.

Band B in Figure 7-8 consists of 24 patterns, representing four instances each of 7sus4, add9, aug7, m (add9), m (maj7), and maj7 tetrachords. Importantly, these are exactly the same types of tetrachords found in Band C of Hidden Unit 2, with one exception: Band B does not include any diminished seventh chords. More importantly, the four instances of each type of tetrachord found in Band B are precisely the four instances not found in Band C of Hidden Unit 2.

What do all of the tetrachords in Band B have in common? Each chord includes two pitch-classes associated with a small negative weight, one pitch-class associated with a strong negative weight, and one pitch-class associated with a strong positive weight. Variation in the weights (for instance, the small negative weight could be either −0.12 or −0.33) produces variation in net input, which is why Band B is wide. On average, a pattern that belongs to this band generates a net input of −0.18, which is close enough to μ to produce strong activity in Hidden Unit 4.

Why does this band capture a different subset of tetrachord instances when Hidden Unit 4 and Hidden Unit 2 organize input pitch-classes according to the same strange circles? Compare the weights in Figure 7-8 to those in Figure 7-7. Note that different weight values are assigned to the same strange circles in the two hidden units. For instance, Hidden Unit 2 assigns a strong positive weight to pitch-classes that belong to the first circle of minor thirds, while Hidden Unit 4 assigns a weak negative weight to the same pitch-classes. These differences cause some instances of a tetrachord type to generate strong activity in one hidden unit, but also to generate weak activity in the other.

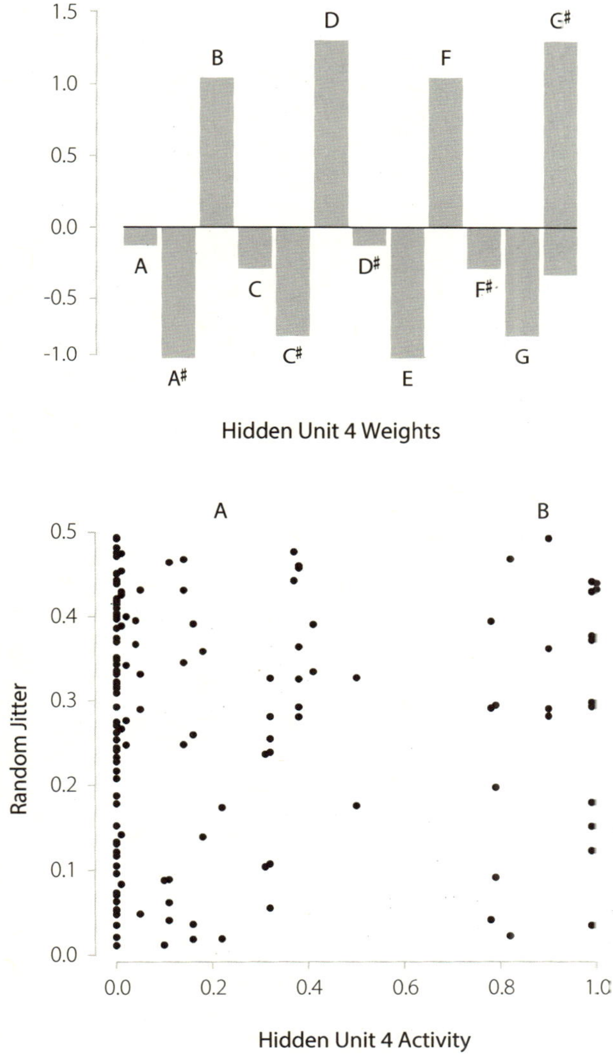

Figure 7-8 The connection weights and the jittered density plot for Hidden Unit 4.

What about Band A in Figure 7-8? None of these patterns is defined with the same combination of pitch-classes (two small negative weights, one large negative weight, and one large positive weight) that signals membership in Band B. Of course, some other combinations of weights produce moderate Hidden Unit 4 activity, but none is as optimal as the Band B combination. High activity in Hidden Unit 4 represents the detection of this particular combination, which serves to capture 24 tetrachords that (musically) should have been in Band C of Hidden Unit 2, but were not.

7.3.5 Hidden Unit 7

Figure 7-9 presents the connection weights and the jittered density plot for Hidden Unit 7 of the extended tetrachord network. Importantly, at the end of training the value of μ for this hidden unit was −0.02. Thus in order for this unit to generate high activity, the four signals being sent to it from input units must cancel each other out to provide a near-zero net input.

The connection weights for this network indicate that it organizes input pitch-classes into equivalence classes defined by the four circles of major thirds. All pitch-classes that belong to the first circle of major thirds have a strong negative weight; those that belong to the second have a weak positive weight; those that belong to the third have a strong positive weight; and those that belong to the fourth have a weak negative weight.

In addition to organizing pitch-classes in terms of circles of major thirds, the connection weight values of Hidden Unit 7 provide an interesting balancing of pairs of pitch-classes. Pairs of pitch-classes that are a major second (e.g., A, B), a tritone (e.g., A, D♯), or a minor seventh (e.g., A, G) apart are balanced, because they are assigned weights that are equal in magnitude but opposite in sign. Pairs of pitch-classes separated by any other musical interval will not cancel each other's signal out because of differences in magnitude or sign of their respective connection weights.

As was the case for Hidden Units 1 and 2, the jittered density plot for Hidden Unit 7 is organized into three distinct bands. Two of these bands (Band A and Band B in Figure 7-9) are associated with low activity in Hidden Unit 7, while patterns that belong to Band C turn Hidden Unit 7 on. Band C in Hidden Unit 7's jittered density plot contains 36 input patterns that comprise all 12 instances of just three different types of tetrachords: 7flat5, 7sus4, and dim7. What do these three different types of chords have in common?

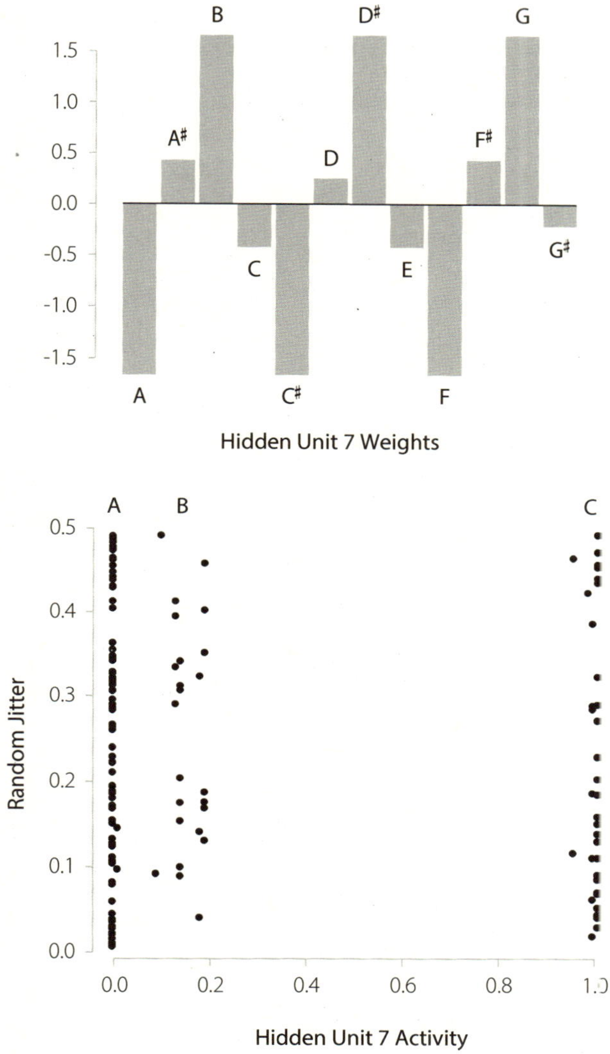

Figure 7-9 The connection weights and the jittered density plot for Hidden Unit 7.

All three of these different types of tetrachords include four pitch-classes that are completely balanced because pairs of these pitch-classes are separated by a major second, a tritone, or a minor seventh. For instance, a diminished seventh chord is composed of two pairs of pitch-classes that are both a tritone apart (Figure 7-2). The two pitch-classes in each pair cancel each other's signal out, producing a net input of zero, which turns Hidden Unit 7 on. Similarly, a 7sus4 chord can be described as two pairs of pitch-classes with each pair separated by a major second (Figure 7-2). As well, a 7flat5 can be described either as two pairs of pitch-classes with each pair separated by a major second, or as two pairs of pitch-classes with each pair separated by a tritone (Figure 7-2). These different descriptions amount to the same effect: two balanced signals from each pair of pitch-classes, generating near-zero net input and turning Hidden Unit 7 on.

The balancing described above is not true of any of the patterns that belong to the other two bands in Figure 7-9. Band B consists of 24 different input patterns comprised of six instances each of 6, aug7, m (add9), and m7 tetrachords. Each of these patterns generates small activity in Hidden Unit 7 (ranging between 0.09 and 0.19) because each is partially balanced in the sense described above. That is, each of these tetrachords contains one pair of pitch-classes that balance because they are separated by a major second, a tritone, or a minor seventh. However, the other pair of tones is not balanced. Interestingly for each of these chords the balanced pair of pitch-classes always involves one weight that is an extreme negative and one that is an extreme positive. They produce some activity in Hidden Unit 7 because the unbalanced pitch-classes involve smaller weights, making net input slightly less than for the remaining tetrachords.

The remaining tetrachords all belong to Band A in Figure 7-9 and all fail to exhibit the kind of balancing discussed above. Eighty-four different patterns belong to this band. Sixty are completely unbalanced tetrachords: a major second, a tritone, or a minor seventh separates none of their pitch-classes. The remaining 24 are the "cousins" of those that belong to Band B. That is, one of their pitch-class pairs is balanced, but the other is not. The difference between these 24 patterns and the 24 that belong to Band B is that they all involve balancing of a weakly negative and a weakly positive weight. As a result, their unbalanced weights are both either extremely positive or extremely negative. As a result, these chords generate an extreme net input, which turns Hidden Unit 7 off.

7.3.6 Hidden Unit 6

Figure 7-10 provides the connection weights and the jittered density plot for Hidden Unit 6. At the end of training, this hidden unit has a value of μ equal to 0.07. The weights presented in this figure indicate that this hidden unit groups pitch-class inputs into equivalence classes based upon the six different circles of tritones. That is, pairs of pitch-classes that are a tritone apart have the same connection weight.

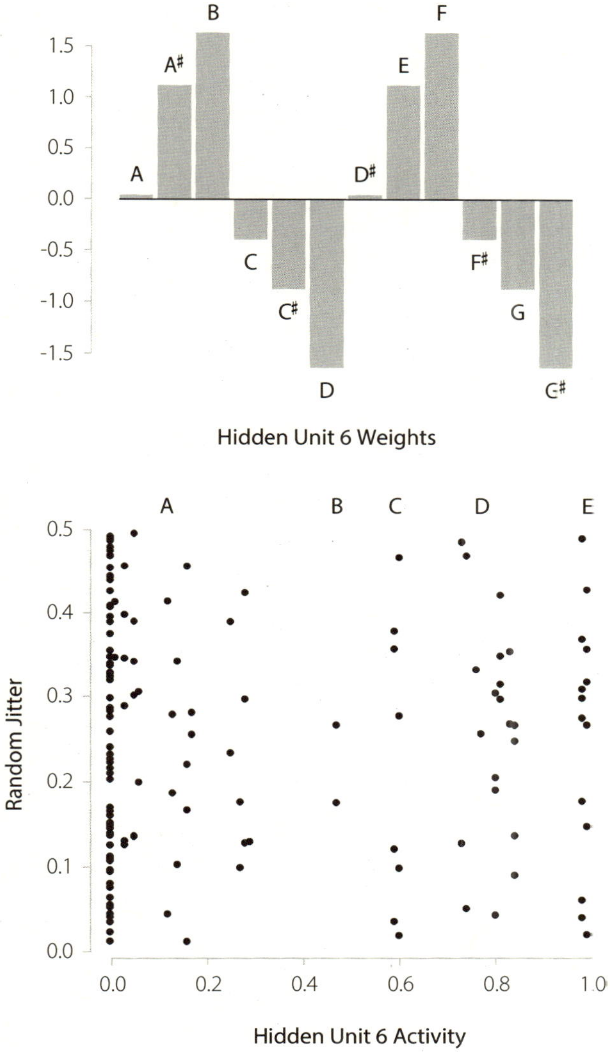

Figure 7-10 The connection weights and the jittered density plot for Hidden Unit 6.

It is also obvious from the weights illustrated in Figure 7-10 that this hidden unit appears to balance, or nearly balance, adjacent triplets of pitch-classes. For instance, consider the first three pitch-classes (A, A♯, B). The pattern of weights assigned to these three inputs seems nearly identical in magnitude but opposite in sign to the pattern of weights assigned to the next three pitch-classes (C, C♯, D) or to the last three pitch-classes (F♯, G, G♯).

Table 7-2 below provides a more accurate indication of which pairs of input pitch-classes cancel each other out given the particular connection weights in Figure 7-10. It is created by only turning on two of the input units that feed into Hidden Unit 6 at a time. The resulting net input is simply the sum of the weights associated with each of the activated input units.

Table 7-2 The activity produced in Hidden Unit 6 by all possible pairs of different input pitch-classes.

	A	A#	B	C	C#	D	D#	E	F	F#	G	G#
A	—	0.03	0.00	0.58	0.05	0.00	1.00	0.03	0.00	0.58	0.05	0.00
A#	0.03	—	0.00	0.28	0.98	0.27	0.03	0.00	0.00	0.29	0.98	0.27
B	0.00	0.00	—	0.01	0.26	0.99	0.00	0.00	0.00	0.01	0.26	0.99
C	0.58	0.28	0.01	—	0.00	0.00	0.58	0.28	0.01	0.10	0.00	0.00
C#	0.05	0.98	0.26	0.00	—	0.00	0.05	0.98	0.26	0.00	0.00	0.00
D	0.00	0.27	0.99	0.00	0.00	—	0.00	0.28	0.99	0.00	0.00	0.00
D#	1.00	0.03	0.00	0.58	0.05	0.00	—	0.03	0.00	0.58	0.05	0.00
E	0.03	0.00	0.00	0.28	0.98	0.28	0.03	—	0.00	0.28	0.98	0.28
F	0.00	0.00	0.00	0.01	0.26	0.99	0.00	0.00	—	0.01	0.26	0.99
F#	0.58	0.29	0.01	0.10	0.00	0.00	0.58	0.28	0.01	—	0.00	0.00
G	0.05	0.98	0.26	0.00	0.00	0.00	0.05	0.98	0.26	0.00	—	0.00
G#	0.00	0.27	0.99	0.00	0.00	0.00	0.00	0.28	0.99	0.00	0.00	—

Note. For a particular activity in the table, the pitch-class label for the row provides one member of the pair, and the pitch-class label for the column provides the other member. Pairs that cancel each other's signal out, producing high activity in the hidden unit, are indicated by the dark grey cells. Pairs that weakly cancel each other out, producing moderate activity, are indicated by the lighter grey cells.

Each net input in this table can be fed into a Gaussian activation function (with $\mu = 0.07$) to determine the activity produced in Hidden Unit 6. This activity is reported in each cell in Table 7-2. In this table, the column label indicates one of the activated input units, and the row label indicates the other. (Pairs that correspond to the diagonal of the matrix were not presented, because in this multilayer perceptron it is not possible to send two signals simultaneously from one input unit.) If the Gaussian activity produced is 0.90 or higher, then this indicates that the signals from the two input units cancel each other out, turning Hidden Unit 6 on. The input pairs that cancel each other out have grey cells in Table 7-2.

If tritone balancing were the only kind of balancing evident in Table 7-2, then only six different pairs of pitch-classes would cancel each other's signal out. An inspection of Table 7-2 indicates that nine different pairs of inputs cancel one another out. (Note that the table is symmetric, and that each pair occurs twice in the table.) Each is highlighted in a dark grey cell in the table. In addition, four other pairs of pitch-classes nearly cancel one another out, because they produce activity of 0.58. The weaker activity produced by these pairs of inputs is highlighted with lighter grey cells in the table. The pattern of grey cells in Table 7-2 is very regular, consistent with the regular pattern of alternating connection weights in Figure 7-10. In general, pitch-class pairs that are separated by a minor third or by a major sixth cancel one another out. There are two caveats to add to this general description. First, in some instances (e.g., A paired with C) the two weights are different enough in magnitude that they do not completely cancel one another out, but cancel each other out enough to produce moderate activity. Second, even though A and D♯ are a tritone apart, their connection weights are so close to zero that this pair produces high activity in Hidden Unit 6 too.

With this understanding of the connection weight structure in Figure 7-10, let us now consider the nature of the bands in the jittered density plot for Hidden Unit 6.

The jittered density plot for Hidden Unit 6 reveals five different bands. Excluding Band A (which again appears to be a "zero loading" band with no interpretable structure), these bands share one interesting qualitative characteristic: all of the tetrachords that belong to the same band are missing a pair of pitch-classes. Patterns in Band E are missing both A and D♯; patterns in Band D are missing both D and G♯. Each of these missing pairs defines a tritone circle (i.e., [A, D♯] or [D, G♯]). Patterns in Band C are all missing both A♯ and G, which are separated by a minor third. The two patterns that belong to Band B (C♯m6 and Gm6) are missing A and D♯, B and F, and C and F♯. Each of these pairs defines a tritone circle.

Quantitatively all of the bands in the Figure 7-10 jittered density plot can be explained in terms of the balancing of adjacent pitch-classes. Let us use the connection weights in Figure 7-10 to identify four different sets of three pitch-classes: let Subset 1 be [A, A♯, B], let Subset 2 be [C, C♯, D], let Subset 3 be [D♯, E, F], and let

Subset 4 be [F♯, G, G♯]. Our previous discussion of the connection weights for each of these subsets (see Figure 7-10 and Table 7-2) suggested that if the same pattern of input activity is present in two of these subsets, then their activities will all cancel out, producing high activity. For instance, imagine an input pattern that includes both A and B as pitch-classes. This corresponds to the pattern of activity [1, 0, 1] in Subset 1. If this same pattern of activity is present in Subset 2 or Subset 4, then the signals from the two different subsets will cancel out, producing high activity in Hidden Unit 6. However, if this same pattern of activity is present in Subset 3, the activities will not cancel out, because these two subsets of pitch-classes have the same connection weights.

We can analyze each input-pattern that belongs to a band in terms of the patterns of activity present in each of the four pitch-class subsets for that pattern. We can perform this analysis both qualitatively (e.g., is the pattern of activity in Subset 1 the same as the pattern in Subset 2) and quantitatively (e.g., what is the contribution to net input from Subset 1 or from Subset 2). These analyses indicate that band membership can be explained by patterns of activity balancing across the four different subsets. For example, consider Band E in Figure 7-10. It consists of 14 different tetrachords, including dim7, aug7, m7, and 6 chords. All but two of these input patterns are completely balanced in the sense that they have the same pattern of activity in both Subsets 1 and 2, and have the same pattern of activity in both Subsets 3 and 4. This produces net inputs near 0.07, producing high activity in Hidden Unit 6.

The only exceptions to this are the two augmented seventh chords (F♯aug7 and Caug7) found in Band E. These two tetrachords have identical patterns of activity in Subsets 1 and 3, which do not balance, and which produce a net input of 1.09 from each subset. However, they also have patterns of activity that produce a net input of –0.39 from a third subset, and a net input of –1.66 from the fourth. When all four net input components are combined, the final net input for both chords is 0.13, producing Hidden Unit 6 activity of 0.99.

Band D in Figure 7-10 contains 20 different input patterns, representing a variety of different types of tetrachord [dim7, aug7, m7, 6, m6, add9, and m (maj7)]. Of these 20 patterns, 12 are similar to those described for Band E: Subsets 1 and 2 have the same pattern, as do Subsets 3 and 4. However, because the tetrachords in this band include A and D♯, these two pitch-classes do not completely cancel out corresponding pitch-classes in the other subsets (see Table 7-2). As a result, the net inputs for these patterns are slightly larger, producing slightly lower Hidden Unit 6 activities. This is true even when the patterns of activities in complementary subsets are identical.

The remaining eight patterns in Band D have less balance between subsets but still produce small enough net inputs to generate high Hidden Unit 6 activity. Two of these chords are augmented sevenths that include either an A or a D♯ (Aaug7, D♯aug7). Their patterns of activity across subsets are similar to the two augmented seventh chords in Band E, but their net input is slightly more extreme (around −0.17) because A or D♯ are involved with weaker balance (Table 7-2). The remaining six input patterns in this band involve balance between two of the subsets, but the other two are not balanced. Again, the weights of the particular pitch-classes involved are such that net input is low enough to generate strong activity in Hidden Unit 6.

The remaining bands in the Figure 7-10 jittered density plot involve less balance between subsets and more extreme net inputs, decreasing Hidden Unit 6 activity even further. For instance, Band C consists of eight tetrachords, half of which are sixths and half of which are minor sevenths. None of the subsets balances any of the others for any of these input patterns. However, each of these eight tetrachords has one subset that has a zero net input. The net inputs from the remaining three subsets sums to either −0.34 or 0.34, producing activity of about 0.60

Band B consists of only two tetrachords, C♯m6 and Gm6. Both of these tetrachords have the same pattern of activity in Subsets 1 and 3, producing net input of 1.09 in each. This is the same situation we observed for the two augmented seventh chords that belong to Band E. The difference emerges in terms of the net inputs produced for these two minor sixth chords for the other two subsets, which are −1.66 and −0.94 respectively. In sum, these two chords generate a net input of −0.42, which results in only moderate Hidden Unit 6 activity.

The remaining 98 tetrachords belong to Band A. These are instances of nine of the 12 different types of tetrachord, including all of the 7flat5, 7sus4, m (add9), and maj7 chords. Only the 6, m7, and dim7 tetrachords are not found in this band. In general, there is less and less balance among the four subsets of inputs as one inspects the chords that belong to this band. When balance does occur, it is typically between only two of the subsets; the remaining two subsets are so unbalanced that extreme net input is the result. The net inputs found for the patterns in this band range from −4.11 to 3.43. There is substantial variability in this range, and sometimes net input is small (e.g., around −0.57. This explains why this band is moderately broad in Figure 7-10.

7.3.7 Hidden Unit 5

Figure 7-11 provides the connection weights and the jittered density plot for Hidden Unit 5. At the end of training, its μ is equal to −0.03.

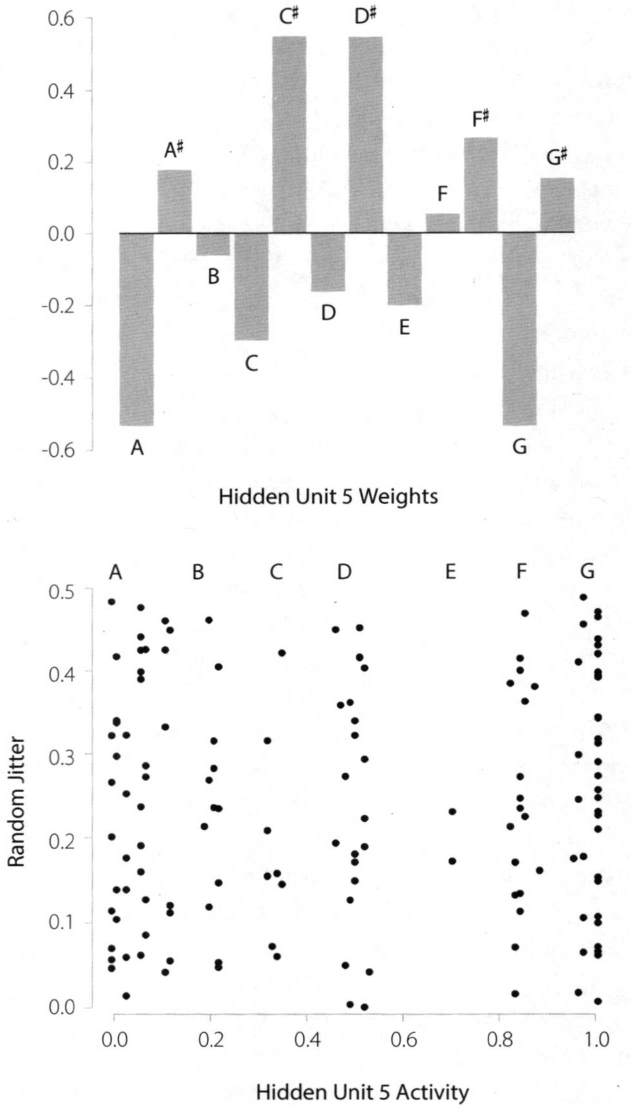

Figure 7-11 The connection weights and the jittered density plot for Hidden Unit 5.

Unlike the previous hidden units that I have analyzed, Hidden Unit 5 does not appear to organize pitch-classes into equivalence classes based upon musical intervals.

Instead, it exhibits tritone balance: pairs of pitch-classes that are a tritone apart have weights that are equal in magnitude but opposite in sign.

Although it is less evident than was the case in Figure 7-10, Figure 7-11 indicates that Hidden Unit 5 is also structured to produce balance between patterns of activity defined over subsets of three adjacent input pitch-classes. Again, let Subset 1 be [A, A♯, B], let Subset 2 be [C, C♯, D], let Subset 3 be [D♯, E, F], and let Subset 4 be [F♯, G, G♯]. An inspection of Figure 7-11's connection weights indicates that two pairs of these subsets appear to balance one another: Subset 3 balances Subset 1, while Subset 4 balances Subset 2.

A quantitative examination of this pattern of connection weights reveals a tremendous amount of balancing or near balancing within its structure. As was done with Hidden Unit 5, I present every possible pair of input pitch-classes to this hidden unit. The net input for each pair is the sum of the weights of the two pitch-classes. I then compute the activity produced in Hidden Unit 5 by passing each of these net inputs through a Gaussian activation function (with $\mu = -0.03$). Table 7-3 presents the results.

Table 7-3 indicates that there is a great deal more balancing possible with the set of connection weights for Hidden Unit 5 than there was for Hidden Unit 6. There are 19 different pairs of pitch-classes that generate activity of 0.9 or higher, indicating near-perfect balance. These cells are highlighted in grey in the table. (Again, this table is symmetric, so that each of these pairs is represented twice.) An additional 28 different pairs of pitch-classes nearly balance, and generate activity that ranges between 0.5 and 0.9.

With this degree of balancing and near balancing between pairs of connection weights, and with the potential for balancing between pairs of subsets of input patterns, it is perhaps not surprising that the jittered density plot in Figure 7-11 exhibits a large number of fairly narrow bands. In order to understand the nature of this banding, we can examine Hidden Unit 5 in terms of the relationships between patterns of activity among the four different subsets of input pitch-classes. Again, this analysis is both qualitative (do the subsets have the same input pattern) and quantitative (what is the net input generated by each subset).

Perhaps not surprisingly, the account of banding for Hidden Unit 5 is very similar to the account detailed for Hidden Unit 6 in the preceding section. For the 40 input patterns that belong to Band G, two different situations emerge. In one, the pattern for both Subsets 1 and 3 is identical, as is the pattern for both Subsets 2 and 4. As a result, near-perfect balance is achieved and Hidden Unit 5 turns on. In the other, the patterns in the various subsets do not balance. However, specific pairs of pitch-classes—from the large number available given Table 7-3—are combined to balance, again turning this hidden unit on.

Table 7-3 The activity produced in Hidden Unit 5 by all possible pairs of different input pitch-classes.

	A	A#	B	C	C#	D	D#	E	F	F#	G	G#
A	—	0.72	0.37	0.14	1.00	0.23	1.00	0.21	0.51	0.85	0.03	0.68
A#	0.72	—	0.93	0.98	0.18	1.00	0.18	1.00	0.81	0.47	0.71	0.65
B	0.37	0.93	—	0.73	0.45	0.88	0.46	0.86	1.00	0.83	0.36	0.95
C	0.14	0.98	0.73	—	0.79	0.55	0.80	0.53	0.87	1.00	0.13	0.97
C#	1.00	0.18	0.45	0.79	—	0.63	0.03	0.66	0.32	0.11	1.00	0.20
D	0.23	1.00	0.88	0.55	0.63	—	0.64	0.69	0.97	0.95	0.22	1.00
D#	1.00	0.18	0.46	0.80	0.03	0.64	—	0.67	0.33	0.12	1.00	0.21
E	0.21	1.00	0.86	0.53	0.66	0.69	0.67	—	0.95	0.96	0.21	1.00
F	0.51	0.81	1.00	0.87	0.32	0.97	0.33	0.95	—	0.69	0.50	0.85
F#	0.85	0.47	0.83	1.00	0.11	0.95	0.12	0.96	0.69	—	0.83	0.51
G	0.03	0.71	0.36	0.13	1.00	0.22	1.00	0.21	0.50	0.83	—	0.67
G#	0.68	0.65	0.95	0.97	0.20	1.00	0.21	1.00	0.85	0.51	0.67	—

Note. Pairs that cancel each other's signal out, producing high activity in the hidden unit, are indicated by the grey cells.

Proceeding through the various bands associated with less activity in Hidden Unit 5, the general story that emerges is the same as that for Hidden Unit 6: there is a growing imbalance between the various pitch-classes that are combined in the patterns that belong to a band, producing more extreme net inputs and lower activity in Hidden Unit 5.

There are some interesting parallels between the contents of some of the bands in Figure 7-10 and the contents of some of the bands in Figure 7-11. For example, Band E for Hidden Unit 6 contains only two augmented seventh chords; another two augmented seventh chords are the only members of Band B for that unit. For Hidden Unit 5, Band G contains only two m (maj7) chords; the two patterns that belong to Band E of the Figure 7-11 jittered density plot are also chords of this type.

Another similarity is that almost all of the bands for Hidden Unit 5 include a diversity of tetrachord types. Indeed, this property seems to be true of almost all of the bands for each of the hidden units for the extended tetrachord network. This property—as well as a detailed listing of the tetrachord types in each band—will be the subject of Section 7.4 later in this chapter.

One difference between the bands for Hidden Unit 5 and the bands for Hidden Unit 6 is that the former set does not contain patterns defined by the absence of specific pairs of pitch-classes. This property is a consequence of the specific patterns of connection weights, and their possible balances, associated with each hidden unit.

7.3.8 Hidden Unit 3

Figure 7-12 provides the connection weights and the jittered density plot for Hidden Unit 3. The connection weights in Figure 7-12 indicate that, like Hidden Unit 5, it exhibits tritone balance. Furthermore, if we consider the weights in terms of the same four subsets that have been applied to the previous two hidden units, Subset 1 balances Subset 3, and Subset 2 balances Subset 4. This pattern was previously observed in Figure 7-11.

At the end of training, the value of μ for this hidden unit is −0.01. I again compute the activity generated by every possible pair of input pitch-classes. The results are shown below in Table 7-4. Table 7-4 indicates that there are 15 different pairs of inputs that cancel one another out perfectly (each pair is represented twice in the table). These pairs produce activity of 0.99 or higher, and have their corresponding cells highlighted in grey in the table. In addition to these pairs, there are 30 other pairs that when combined nearly balance each other's signal, producing activity in Hidden Unit 3 that ranges between 0.5 and 0.9.

It is particularly interesting to compare the pattern of connection weights in Hidden Unit 3 (Figure 7-12) to those for Hidden Unit 5 (Figure 7-11). At first glance, the two patterns seem very similar. However, a closer inspection reveals important differences between the two. Consider the weights for Subset 1 (A, A♯, B) in Figure 7-12, which has a strong negative followed by a moderate positive and a weak negative. This pattern is also evident in Figure 7-11—but for Subset 4 (F♯, G, G♯). Similarly the pattern for Subset 2 in Figure 7-12 is found instead for Subset 1 in Figure 7-11; the pattern for Subset 3 in Figure 7-12 is found for Subset 2 in Figure 7-11; and the pattern for Subset 4 in Figure 7-12 is found for Subset 3 in Figure 7-11. In short, it would appear that both Hidden Units 3 and 5 use the same patterns of connection weights (defined over the four subsets of input units), but assign these same patterns to different subsets of input pitch-classes.

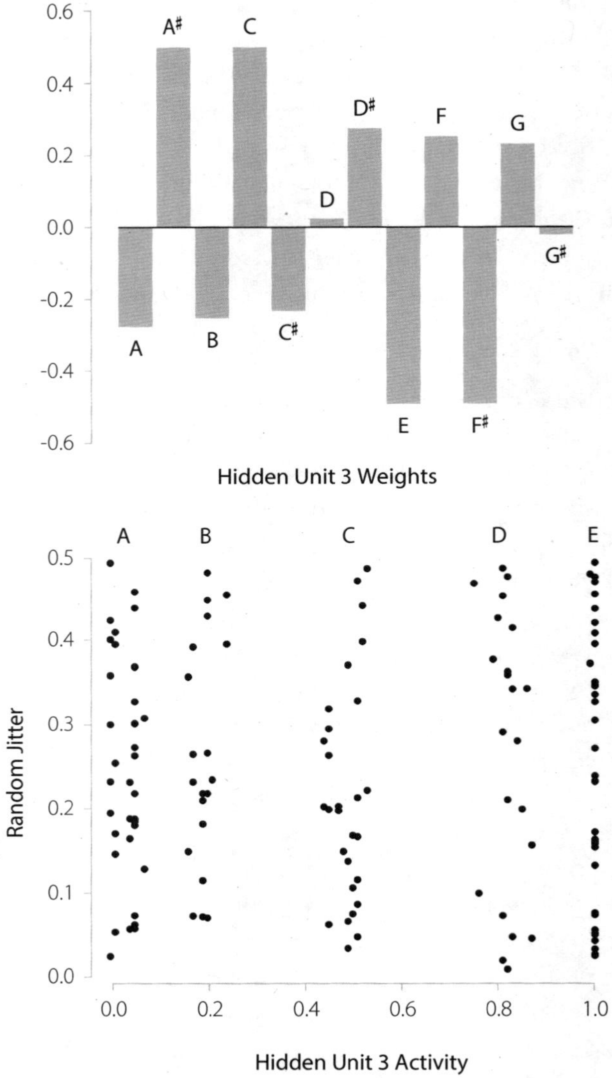

Figure 7-12 The connection weights and the jittered density plot for Hidden Unit 3.

This raises the question: What is the relationship between the responses of Hidden Units 3 and 5 to the set of input patterns, given that there are both interesting similarities and differences between their patterns of connection weights?

Table 7-4 The activity produced in Hidden Unit 2 by all possible pairs of different input pitch-classes.

	A	A#	B	C	C#	D	D#	E	F	F#	G	G#
A	—	0.87	0.44	0.86	0.46	0.83	1.00	0.17	1.00	0.17	0.99	0.76
A#	0.87	—	0.83	0.05	0.81	0.43	0.16	1.00	0.19	1.00	0.20	0.52
B	0.44	0.83	—	0.82	0.51	0.88	0.99	0.21	1.00	0.20	1.00	0.81
C	0.86	0.05	0.82	—	0.80	0.42	0.15	1.00	0.19	1.00	0.19	0.50
C#	0.46	0.81	0.51	0.80	—	0.89	0.99	0.22	1.00	0.21	1.00	0.82
D	0.83	0.43	0.88	0.42	0.89	—	0.73	0.55	0.79	0.53	0.80	1.00
D#	1.00	0.16	0.99	0.15	0.99	0.73	—	0.89	0.42	0.88	0.44	0.81
E	0.17	1.00	0.21	1.00	0.22	0.55	0.89	—	0.85	0.06	0.84	0.46
F	1.00	0.19	1.00	0.19	1.00	0.79	0.42	0.85	—	0.83	0.49	0.86
F#	0.17	1.00	0.20	1.00	0.21	0.53	0.88	0.06	0.83	—	0.82	0.45
G	0.99	0.20	1.00	0.19	1.00	0.80	0.44	0.84	0.49	0.82	—	0.87
G#	0.76	0.52	0.81	0.50	0.82	1.00	0.81	0.46	0.86	0.45	0.87	—

Note. Pairs that cancel each other's signal out, producing high activity in the hidden unit, are indicated by the grey cells.

To answer this question, we can correlate the activities of these two hidden units produced by each of the entire set of input patterns. We can also correlate the activities of each of these units with the activities of Hidden Unit 6. This is because, like the other two, it is sensitive to tritones, and it groups input signals into four different subsets.

Table 7-5 provides the resulting correlations. This table reveals very low correlations between the activities of different hidden units. This means that while there are definite similarities among these units in terms of general patterns of connectivity, their connection weights are arranged in different orders. As a result, the tetrachords that cause high activity in one hidden unit do not do so for the other two. This will be important in our consideration of coarse coding in the next section of this chapter.

Table 7-5 Correlations among activities of three hidden units to the 144 input patterns.

	HID3	HID5	HID6
HID3	1.00		
HID5	0.11	1.00	
HID6	0.00	0.02	1.00

Given that Hidden Unit 3 uses patterns of connection weights similar to those found in Hidden Units 5 and 6, but assigns the weights to different subsets of inputs, I expect that an account of the various bands in Figure 7-12 would be very similar to the accounts provided earlier for both Hidden Unit 5 and Hidden Unit 6. This is indeed the case. For the 38 input patterns that belong to Band G, two different situations emerge. For 24 of the patterns in this band, the pattern of activity for both Subsets 1 and 3 is identical, as is the pattern for both Subsets 2 and 4. As a result, near-perfect balance is achieved and Hidden Unit 3 turns on. In the other, the patterns in the various subsets do not balance. However, specific pairs of pitch-classes (for the potential, see Table 7-4) combine to balance, again turning this hidden unit on.

Proceeding through the various bands associated with less activity in Hidden Unit 3, the general story that emerges is the same as for Hidden Units 5 and 6: there is a growing imbalance between the various pitch-classes that are combined in the patterns that belong to a band, producing more extreme net inputs and lower activity in Hidden Unit 3.

Of course, each of the bands in Figure 7-12 picks out a variety of different tetrachords. A detailed list of those for the various bands in Hidden Unit 3's jittered density plot is presented in the next section's discussion of coarse coding in the extended tetrachord network.

7.4 Bands and Coarse Coding

7.4.1 Hidden Unit Structure

Section 7.3 presented a detailed examination of the structure of each of the seven hidden units in the extended tetrachord network. This revealed many details about the connection weight structure of each hidden unit, as well as the types of tetrachords that produced varying degrees of activity in each hidden unit. These details reveal three general points. First, the connection weight structure of each hidden unit is highly regular, and this structure relates to musical intervals. Four of the hidden units assign input pitch-classes to equivalence classes based upon strange circles. Hidden Unit 1 groups pitch-classes using the two circles of major seconds and the six circles of tritones. Hidden Unit 2 assigns pitch-classes to equivalence classes based upon the three circles of major thirds. Hidden Unit 4 organizes pitch-classes using both the three circles of minor thirds and the six circles of tritones. Hidden Unit 6 uses equivalence classes based on circles of tritones. Hidden Unit 7 assigns pitch-classes to equivalence classes based upon the four circles of major thirds. The remaining two hidden units (Hidden Units 3 and 5) employ tritone balance, assigning weights that are equal in magnitude but opposite in sign to pitch-classes separated by a tritone.

Second, the connection weight structure of each hidden unit produces distinct banding when hidden unit activities are graphed in a jittered density plot. The hidden units that organize pitch-classes using strange circles exhibit either two or three distinct bands, while the hidden units that employ tritone balance exhibit five or six distinct bands. For all hidden units these bands emerge because signals from different pairs of pitch-classes are assigned connection weights that produce balance or near balance for some pairs but not others.

Third, almost all of the bands in each hidden unit's jittered density plot are heterogeneous. That is, almost every band includes instances of more than one type of tetrachord. This is apparent in Table 7-6, which lists each tetrachord type found in each band of the seven jittered density plots.

Table 7-6 The types of tetrachords found in each band in each jittered density plot that was presented in Section 7.3.

Hidden unit	Band	Tetrachords in band
1	A	aug7, 7flat5
	B	7, m6, m(maj7), add9
	C	6, m7, maj7, dim7, 7sus4, m(add9)
2	A	6, 7, m6, m7, maj7, dim7, aug7, m(maj7), 7flat5, 7sus4, add9, m(add9)
	B	6, 7, m6, m7, 7flat5
	C	maj7, dim7, aug7, m(maj7), 7sus4, add9, m(add9)

Hidden unit	Band	Tetrachords in band
3	A	6, m7, maj7, 7sus4, add9, m(add9)
	B	7, m6, m(maj7), add9
	C	6, m7, maj7, aug7, 7sus4, m(add9)
	D	7, m6, m(maj7), add9
	E	6, m7, maj7, dim7, aug7, 7flat5
4	A	6, 7, m6, m7, maj7, dim7, aug7, m(maj7), 7flat5, 7sus4, add9, m(add9)
	B	6, 7, m7, maj7, 7flat5, 7sus4, m(add9)
	C	maj7, aug7, m(maj7), 7sus4, add9, m(add9)
5	A	6, 7, m6, m7, maj7, m(maj7), 7sus4, add9, m(add9)
	B	7, m6
	C	aug7, 7sus4, add9, m(add9)
	D	7, m7, maj7, aug7, 7sus4, m(add9)
	E	m(maj7)
	F	7, m6, m(maj7), add9
	G	6, m7, maj7, dim7, aug7, m(maj7), 7flat5
6	A	7, m6, maj7, aug7, m(maj7), 7flat5, 7sus4, add9, m(add9)
	B	7, m6, dim7, aug7
	C	7, m(maj7), add9
	D	m6
	E	6, m7
	F	6, m6, m7, dim7, aug7, m(maj7), add9
	G	6, m7, dim7, aug7
7	A	6, 7, m6, m7, maj7, aug7, m(maj7), add9, m(add9)
	B	6, m7, aug7, m(add9)
	C	dim7, 7flat5, 7sus4

Table 7-6 indicates that only four of the 31 different bands are pure in the sense that they pick out only one type of tetrachord. Hidden Unit 5 Band B contains only m6 chords (which are identical to the 7 chords that it also contains). Hidden Unit 5 Band E contains only m (maj7) chords. Hidden Unit 6 Band D only contains m6 chords; and Hidden Unit 6 Band E contains only 6 chords (which are identical to the m7 chords that it also contains). Every other band contains at least two different types of tetrachords. Many of these bands contain six or more different types of tetrachords.

Table 7-6 indicates some additional properties concerning the similarities and differences between different bands and their contents. For instance, several different types of chords appear to be similar to one another because they are frequently seen together in the same band. For instance, tetrachords that belong to the four types 7, add9, m (maj7), and m6 are found together in eight of the 31 different bands in Table 7-6.

There are also substantial differences between individual hidden units in terms of their sensitivity to such groups of chords. For instance, bands that contain 7, add9, m (maj7), and m6 chords are associated with different levels of activity when bands from different hidden units are compared (compare Hidden Unit 1 Band B to Hidden Unit 5 Band F). As well, some units that have bands that contain these four chord types contain other tetrachord types as well, and these typically differ from one another. For instance, Hidden Unit 7 Band A groups these four chord types along with instances of 6, maj7, aug7, m (add9), and m7 chords. In contrast, Hidden Unit 4 Band A includes these four types along with instances of 6, m7, aug7, 7♭5, dim7, m (add9), maj7, and 7sus4 tetrachords. Furthermore, these four chord types are not always found in the same band. For example, Hidden Unit 6 Band C contains 7, add9, and m (maj7) chords, but does not contain any m6 tetrachords.

7.4.2 Bands and Coarse Coding

The general summary of network structure provided above indicates two key facts: the hidden units of the extended tetrachord network are highly structured, but individual hidden units do not detect the presence or absence of particular tetrachord types. How then do the output units of the extended tetrachord network process hidden unit activity to identify an input pattern's chord type?

The general answer to this question is that the extended tetrachord network is another example of coarse coding, a concept that was introduced in Chapter 5. In coarse coding, individual hidden units serve as inaccurate detectors of input pattern properties. However, particularly if each hidden unit views the inputs from a different perspective, when different inaccurate representations are combined, an accurate classification emerges. Fortunately, the summary of band contents in Table 7-6 helps provide an explanation of coarse coding in this particular network.

Imagine presenting one input chord to the trained extended tetrachord network and only observing the activity that it produces in each of the seven hidden units. The chord produces the following activity pattern, given in ascending order of hidden units from H1 to H7: [0.14, 0.02, 0.01, 0.99, 0.32, 0.05, 0.00]. Given this activity pattern, what type of tetrachord was presented?

To answer this question, let us re-label each hidden unit activity value with the jittered density plot band to which that activity value corresponds. When this is done, the set of hidden unit bands to which the pattern belongs (in the same order as before) is: [B, A, A, C, C, A, A]. With this pattern of bands in hand, let us take Table 7-6 and delete any bands that are not present in this set. The result is presented as Table 7-7:

Table 7-7 The types of tetrachords found in each band to which the first single pattern presented to the network belongs.

Hidden unit	Band	Tetrachords in band
1	B	7, m6, m(maj7), **add9**
2	A	6, 7, m6, m7, maj7, dim7, aug7, m(maj7), 7flat5, 7sus4, **add9**, m(add9)
3	A	6, m7, maj7, 7sus4, **add9**, m(add9)
4	C	maj7, aug7, m(maj7), 7sus4, **add9**, m(add9)
5	C	7, m6, m(maj7), **add9**
6	A	7, m6, maj7, aug7, m(maj7), 7flat5, 7sus4, **add9**, m(add9)
7	A	6, 7, m6, m7, maj7, aug7, m(maj7), **add9**, m(add9)

Note. The only chord type that is found in each band is the added ninth (add9), which is indicated in bold font.

Note that each band in Table 7-7 is inaccurate, in the sense that it contains four or more types of tetrachords. However, only one tetrachord type is present in all seven of these bands: add9. [The 7, m6, and the m (maj7) are all absent from Hidden Unit 3 Band A, while none of the other chords (apart from add9) are present in Hidden Unit 1 Band B.] This means that the tetrachord presented to the network must have been an add9.

Consider a second example, a chord that produces the following pattern of hidden unit activity: [0.16, 0.24, 0.20, 0.00, 0.12, 0.14, 0.00]. In terms of band labels, this pattern is equivalent to [B, B, B, A, A, B, A].

Table 7-8 The types of tetrachords found in each band to which the second single pattern presented to the network belongs.

Unit	Band	Tetrachords in band										
		7	**add9**	**m(maj7)**	**m6**	**6**	**7flat5**	**m7**	**dim7**	**m(add9)**	**maj7**	**7sus4**
1	B	x	x	x	x							
2	B	x			x	x	x	x				
3	B	x	x	x	x							
4	A	x	x	x	x	x	x	x	x	x	x	x
5	A	x	x	x	x	x		x		x	x	x
6	B	x			x				x			
7	A	x	x	x	x	x		x		x	x	

Note. The only type of chord found in each band is the dominant seventh (7), which is indicated in bold font.

Table 7-8 represents each of these bands in terms of their component tetrachord types. This table indicates that the only tetrachord type found in every band is the dominant seventh. Therefore, the stimulus presented to the network was a 7 chord. The two examples of coarse coding illustrated in Tables 7-7 and 7-8 were chosen deliberately. We noted earlier that in terms of band contents add9 and 7 chords were similar because they are often in the same band. However, the coarse coding examples show that there are indeed differences between the two chord types; there are some bands where we find one but not the other. This band intersection technique takes advantage of this property, which explains how the messy contents of the 31 hidden unit bands permit identification of the type of chord presented to the extended tetrachord network.

The output units of the extended tetrachord network do not themselves literally identify chord types by determining intersections between sets of features captured by different hidden unit activities. Instead, the output units operate geometrically: hidden unit activities provide coordinates that arrange particular types of tetrachords along a plane, and the output units then carve this plane out of the hidden unit space (e.g., Figures 6-31 and 6-32). Functionally speaking, however, this geometric process of identifying tetrachord types is equivalent to finding intersections between bands. Tetrachord types that belong to the same band will have nearly the same coordinate along one of the dimensions of the hidden unit space. One type of tetrachord is separated from the other types in this dimension by being located at a different coordinate from the others in one or more of the other dimensions.

7.5 Summary and Implications

Chapter 7 has provided an account of a network trained on a third version of a chord classification task, identifying members of a set of extended tetrachords. This task is more complicated than those tasks described in Chapter 6 because the training set contained a broader variety of chord types. As a result, a more complex multilayer perceptron, one that used seven hidden value units, was required for this problem.

Interestingly, when I interpreted this more complex network, we discovered some properties that were very similar to those discovered in the networks discussed in Chapter 6. In particular, many of this network's hidden units employed strange circles to solve this problem. Recall that a strange circle picks out 1) pitch-classes that belong to a particular interval cycle and 2) assigns all of the pitch-classes within this subset nearly identical connection weights. In other words, the hidden units use interval cycles to define equivalences between different pitches. Hidden Unit 1 did this with circles of major seconds, Hidden Units 2 and 4 organized inputs using circles of minor thirds, Hidden Unit 6 applied circles of tritones, and Hidden Unit 7 illustrated circles of major thirds. The other two hidden units balanced particular subsets of intervals, but not in a way that reflected perfect use of strange circles.

The interpretation of the extended tetrachord network also provided an opportunity to explore the use of distributed representations. The jittered density plots for each hidden unit revealed that different levels of activity in each unit picked out a particular subset of input patterns. Musically speaking, each of these subsets was very hard to understand. However, by combining the subsets picked out by each of the seven hidden units, and looking for the chord types that belonged to each, it became clear how the network used this distributed representation to solve the extended tetrachord classification problem.

The extended tetrachord network reveals two interesting properties that also appeared in Chapter 6. First, its hidden units have a marked tendency to use a construct from post-tonal music theory—interval cycles—to classify entities that belong to tonal music. Furthermore, it uses these cycles in a strange way, by creating equivalence classes so that pitch-classes that belong to the same interval cycle are all treated as being the same pitch-class. In other words, rather than using the 12 pitch-classes that are the foundation of Western music theory (be it tonal or atonal), this network's hidden units operate in a musical world in which there are fewer than 12 pitch-classes. This is indeed an alien music theory.

This network also reveals an interesting example of an alternative representation for musical cognition, the distributed representation. Is it possible that when human cognition processes musical stimuli similar coarse codes are employed? One way to answer this question would be to explore the kinds of errors made by humans when learning to perform chord classification, or when classifying chords (post-learning) under additional attentional demands. We could compare these to network errors made during learning, or network errors made after particular hidden units are ablated from the network. Similar patterns of errors might point toward the discovery of a coarse code for human musical cognition.

8

Jazz Progression Networks

8.1 The ii-V-I Progression

All of the musical tasks that I have considered in previous chapters ignored the element of time. One reason for this, detailed in Chapter 2, was that the network architectures that I have explored are designed to recognize spatial, but not temporal, patterns. However, even these sorts of networks can be used to deal with some temporal aspects of music. In this chapter, I explore time by training networks on sequences of chords. After learning, if a stimulus chord is presented to the network, then it responds with the next chord in the sequence.

The chapter begins with an introduction to chord progressions, focusing upon one important progression in jazz, the ii-V-I. Various methods are discussed that can be used to encode the chords in this progression for network training. I discover that the type of encoding is an important determinant of how long it takes a network to learn this progression. A network is then interpreted to reveal how it encodes the probabilistic structure of this progression. The chapter then turns to a second progression, the Coltrane changes, which is an elaboration of the ii-V-I progression. Various network encodings are explored, and their impact on learning is determined. The interpretation of this network relates the Coltrane changes to the strange circles that have been discussed in the preceding two chapters.

8.1.1 Chord Progressions

The basic element of harmony is the musical interval, the simultaneous presence of two tones a specific musical distance apart. Chords generally involve presenting more than two tones simultaneously, and therefore involve the presence of several musical intervals. Just as the presence of a single tone cannot by itself establish a musical key, a triad in isolation cannot establish tonality (Schoenberg, 1969). In order for tonality to be established, a succession of triads—or, more generally, a succession of harmonies—must occur. The structure of this succession ensures that one chord naturally leads the listener to the next. Such a structured succession of chords is called a chord progression or, in jazz, "the changes." Chord progressions are central to the structure of most jazz compositions (Broze & Shanahan, 2013; Sudnow, 1978). This chapter explores the properties of networks that learn particular progressions from jazz: when presented a chord, the network responds with the next chord in the progression.

8.1.2 Basic Changes

Let us begin with succession of chords called the ii-V-I chord progression, which is likely the one most commonly encountered in jazz (Levine, 1989). In its typical form, this progression involves three different tetrachords, each defined in the same musical key; as a result, we can write the ii-V-I progression for each of the 12 different major keys in Western music. The three chords in any of these versions of the progression are constructed using particular notes in a major scale as their root; the scale used defines the key of the progression.

The first tetrachord in the ii-V-I progression is the minor seventh chord constructed using the second note of the progression's major scale. This is the ii chord; its Roman numeral is written in lower case because it is minor, and indicates the position of this chord's root in the major scale for the chord's musical key. For instance, the second note in the C major scale is D, so the ii tetrachord for the key of C is Dm7, which includes the notes D, F, A, and C.

The second tetrachord in the ii-V-I progression is the dominant seventh tetrachord constructed using the fifth note of the progression's major scale as its root. In the C major scale this note is G, so in the key of C the V chord in the progression is G7, which uses the notes G, B, D, and F.

The third tetrachord in the ii-V-I progression is the major seventh tetrachord constructed using the first note of the progression's major scale as its root. In the C major scale this note is C, so in the key of C the I chord in the progression is Cmaj7, which contains the notes C, E, G, and B.

The procedure illustrated above for constructing the three chords in the key of C can construct the ii-V-I progression in any other major scale. Table 8-1 provides the three chords in this progression for each major key in Western music.

Table 8-1 The three tetrachords that define the ii-V-I progression for each major key

Key	Chord		
	ii	V	I
A	Bm7	E7	Amaj7
A#	Cm7	F7	A#maj7
B	C#m7	F#7	Bmaj7
C	Dm7	G7	Cmaj7
C#	D#m7	G#7	C#maj7
D	Em7	A7	Dmaj7
D#	Fm7	A#7	D#maj7
E	F#m7	B7	Emaj7
F	Gm7	C7	Fmaj7
F#	G#m7	C#7	F#maj7
G	Am7	Dm7	Gmaj7
G#	A#m7	D#7	G#maj7

Note. Each row provides the three chords in the progression for one musical key. The name of the major key is provided in the first column.

8.1.3 The ii-V-I Progression Problem

I am interested in training networks to generate the ii-V-I progression in any key. When presented one chord, the network's task is to generate a representation of the next chord in the progression. For example, consider the ii-V-I progression in the key of C, which involves the Dmin7, G7, and Cmaj7 chords. I want to train a network so that when Dmin7 is presented to its input units it responds with a representation of G7 in its output units. Similarly, when G7 is presented to its input units, it will generate Cmaj7 in its output units.

I want analogous behaviour from the network for the other 11 possible musical keys. Each key involves defining two input/output pairs, one involving the minor seventh and the dominant seventh chords, the other involving the dominant seventh and the major seventh chords. I never use a major seventh chord as an input pattern; when properly trained the network will never generate a minor seventh chord as a response. The entire training set consists of 24 different input/output pattern pairs.

The input and output chords for the ii-V-I progression problem can be encoded in a number of different ways. Of particular interest in the current chapter is whether the choice of encoding affects the complexity of the network required to learn the progression. Before exploring the results of training networks on the ii-V-I progression problem, let us first discuss the importance of encoding, and how there are several different approaches to encoding tetrachords that are worthy of exploration.

8.2 The Importance of Encodings

8.2.1 Readiness-To-Hand

In *Being and Time* (Heidegger, 1927/1962), philosopher Martin Heidegger proposed that part of an agent's engagement with the world involves using equipment. Equipment consists of entities experienced in terms of the potential actions or experiences that they make available. Heidegger also argued that a key property of equipment was readiness-to-hand. Readiness-to-hand means that equipment itself is imperceptible to us when being used; we experience the effects of equipment but not equipment itself. In other words, if we are aware of the existence of a tool, then the tool is poorly designed (Dourish, 2001; Norman, 1998, 2002, 2004). The invisibility of artifacts—the readiness-to-hand of equipment—provides evidence of good design.

8.2.2 Solutions by Design

Readiness-to-hand is not only relevant to the design of artifacts but is also important to theories of problem solving. In cognitive science, problem solving is typically described as searching a problem space (Newell & Simon, 1972). The amount of time required to search through a problem space to find a route to the problem's solution reflects a problem's difficulty. The longer the search, the harder the problem. Crucially, search complexity depends in part upon the manner in which states of knowledge about the problem are encoded. If a problem is encoded using one representational scheme, then its solution may require a long and difficult search. However, if the same problem is encoded in a different format, then its difficulty can be drastically reduced. With the proper encoding, a problem's solution exhibits readiness-to-hand: the solution is immediately apparent, and the process of searching for the solution is so trivial that it becomes invisible (Simon, 1969).

In this context, one theme of the current chapter is to explore artificial neural networks for jazz progressions in the context of efficient design. In particular, it is possible to use many different encodings of the same musical problem. Even though the musical problem remains constant, changing its encoding can make it much more difficult—or much easier—for a network to learn.

8.3 Four Encodings of the ii-V-I Problem

In designing a training set for teaching a network the ii-V-I progression, one must decide how to represent tetrachords both as stimuli and as responses. Ideally, the choice of representation would be "theory neutral" (Pylyshyn, 1984): regardless of our choice of representation, the results of training a network on the task would be the same. Not surprisingly, though, this ideal situation does not arise: different choices of how to represent tetrachords for the network lead to very different simulation results.

Let us first describe four plausible methods for representing tetrachords to networks that must learn the ii-V-I progression. Later in the chapter, we will present results that clearly show that these choices are not theory neutral.

8.3.1 Pitch-Class Encoding

Most of the networks that are described earlier in this book employ a pitch-class representation, which is the first kind of encoding to consider for the ii-V-I progression. This representation only requires 12 units. Each unit represents the presence or absence of one of the possible pitch-classes in Western music.

One major advantage of pitch-class representation is its simplicity: a very small number of input and output units are required to represent any of the different tetrachords that can occur in the progression. A pitch-class representation of the ii-V-I problem requires only 12 input units to represent an input tetrachord, and the same number of output units to represent the tetrachord response generated by the network.

In pitch-class encoding, as we have seen in earlier chapters, a tetrachord stimulus is represented by turning on the four input units that represent the chord's component pitch-classes, and by turning all of the other eight input units off. For the ii-V-I problem, a network can use the same encoding to represent its tetrachord responses in the output units.

8.3.2 *Pitch Encoding*

One straightforward way to consider the various chords in a progression is to consider them as being in root position. In root position, all notes in a chord appear in their natural positions in the scale to which they belong. For instance, in root position the lowest note of a tetrachord is the chord's root: the lowest note of Dm7 is D; the lowest note of G7 is G, and so on.

One consequence of having each tetrachord in root position is that there is a marked similarity in chord "shape," which is the spacing between adjacent notes in the chord. Tetrachords of the same type (minor seventh, dominant seventh, or major seventh) have very similar shape: four notes that are evenly spaced, as they are stacked upon each other on the staff.

One can imagine that the input units used for pitch-class encoding are the keys of a small piano. Figure 8-1 illustrates the mapping between the input units and the piano keyboard. However, this mapping reveals a possible disadvantage of pitch-class representation: by adopting this encoding, we lose the similarity of shape between different chords of the same type. That is, different spacing between notes—different chord inversions—are required to fit tetrachords on this keyboard because of its small size.

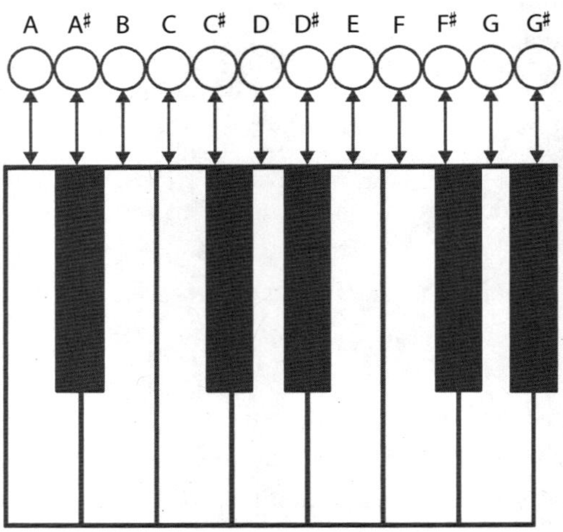

Figure 8-1 The mapping between input units used for pitch-class encoding and a piano keyboard.

Figure 8-2 illustrates this issue using a keyboard to represent four different minor seventh chords that belong to different keys. Each belongs to the ii-V-I progression in a particular key. However, to fit each of these chords onto the small keyboard, different chord shapes are required. For example, Figure 8-2 shows that Amin7 can be fit on this keyboard in root position (the A is the lowest note, which is the leftmost note coloured grey in the illustration). In contrast, Cmin7 must be fit using its first inversion (C is the second lowest note), Dmin7 must be fit using its second inversion (D is the second highest note), and Gmin7 must be fit using its third inversion (G is the highest note).

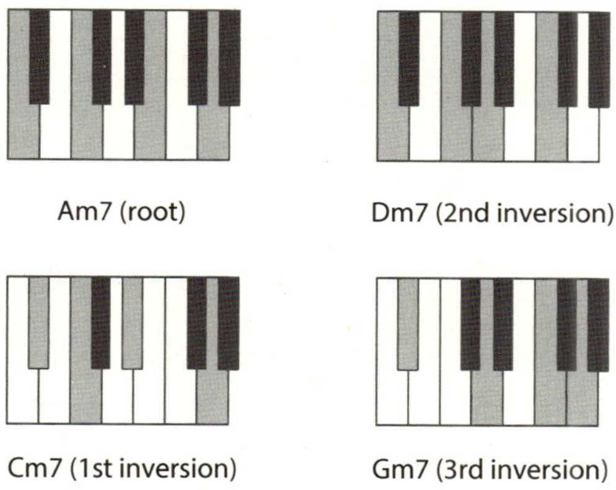

Figure 8-2 The keyboard layout of four different minor seventh tetrachords.

In order to create a representation that preserves tetrachord shape, I must abandon the assumption of octave equivalence, and adopt an encoding that explicitly indicates that two different notes (e.g., middle C and the C an octave higher) are distinct pitches even though they belong to the same pitch-class. I did this for the first triad classification network described in Chapter 6 (see Figure 6-2). Pitch encoding is an alternative to pitch-class encoding, and abandons the octave equivalence assumption. In pitch encoding, each input unit represents the presence or absence of a particular pitch, and not of a pitch-class, as is shown in Figure 8-3.

In our use of pitch encoding for the ii-V-I problem, the highest key of the progression is G♯, and the highest note is C♯6 (the highest note in the D♯7 tetrachord for this key). Similarly, the lowest key of the progression is A, and as a result the lowest note that we used is A3 (the lowest note in the Ama7 tetrachord for this key). As a result, our pitch encoding of chords used 29 input units to represent all of the pitches from A3 to C♯6.

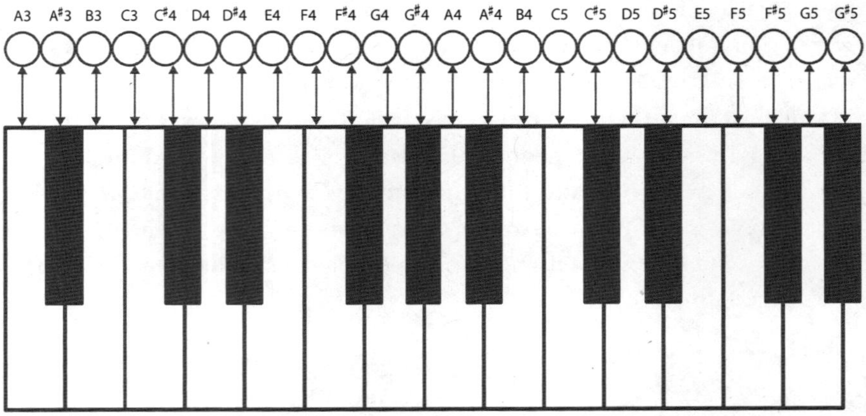

Figure 8-3 The mapping between input units used for pitch encoding and a piano keyboard.

8.3.3 Pitch Encoding of Inversions

Pitch encoding can represent tetrachords in root position. However, as soon as octave equivalence is abandoned, other versions of the ii-V-I progression problem are possible. For instance, a pianist might prefer inversions of the chords that reduce the hand and finger movement required when one moves from one chord to the next. For example, if one uses the second inversion of every dominant seventh chord in the progression, then a "lower action" version of the progression emerges. The second inversion of a dominant seventh chord is created by taking the two lowest notes in the chord's root position and raising each an octave. Figure 8-4 provides a version of the ii-V-I progression in which each of the dominant seventh chords is inverted.

Figure 8-4 The ii-V-I progression for each possible key.

How does inverting the middle chord of the ii-V-I progression enable lower action movement for a pianist? Figure 8-5 illustrates voice leading—that is, finger movements from one chord to the next—for the ii-V-I progression in the key of C to shed light on this issue. The top three keyboards in Figure 8-5 illustrate voice leading when the dominant seventh chord is in root position. The arrows indicate finger movements from chord to chord. Note that because the middle chord is in root position, substantial movement from chord to chord is required: each finger moves to a different key to play the next chord, and the hand must move up and then back down along the keyboard.

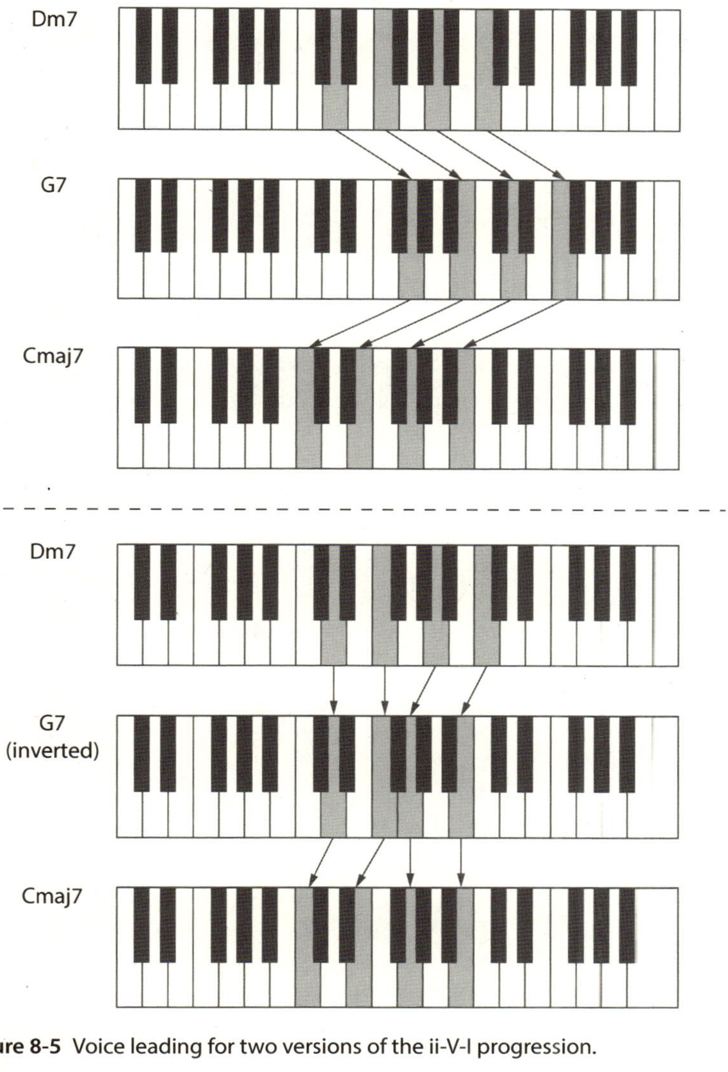

Figure 8-5 Voice leading for two versions of the ii-V-I progression.

The lower half of Figure 8-5 shows that if the middle chord is played in second inversion form, much less movement is required. The hand stays at the same position along the keyboard, and moving from one chord to the next only requires changing the position of two fingers. Two fingers press the same keys in successive chords for this version of the progression! In short, an alternative approach to encoding the ii-V-I progression problem is to use pitch encoding, but also to take advantage of its flexibility by presenting dominant seventh chords in their second inversion form. One consequence of this is that slightly fewer processing units are required; all of the tetrachords can be encoded using 24 input units, with the lowest unit representing A3 and the highest unit representing G♯5.

8.3.4 Lead Sheet Encoding

All of the encodings described above represent each pitch-class or each pitch in a tetrachord. As a result, all involve activating four processing units and turning all of the remaining processors off. However, there are many other ways to represent tetrachords, and some of these representations are not concerned with detailing each note in a chord. For instance, one popular approach to teaching adults how to play piano (Houston, 2004) attempts to simplify music reading by eliminating traditional musical notation of chords. Instead, chords are represented in lead sheet notation: they are written as a combination of the name of one note (to provide the chord's root) and some additional symbols that indicate the type of chord. For instance if one uses lead sheet notation for the ii-V-I progression in the key of C, the chords are merely written as "Dm7," "G7," and "Cmaj7."

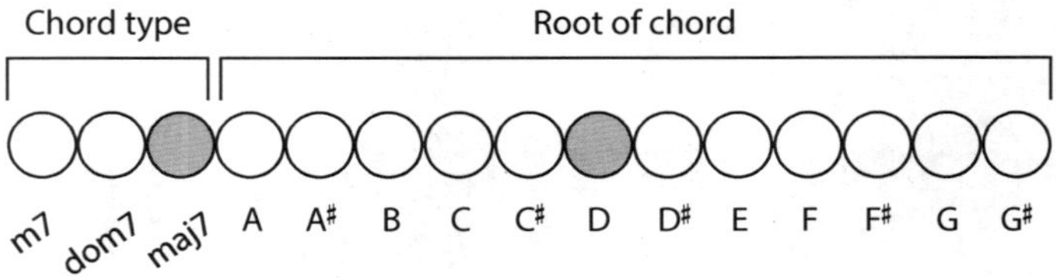

Figure 8-6 Lead sheet encoding of tetrachords.

A lead sheet encoding can be easily created for an artificial neural network that is to learn the ii-V-I progression. This encoding is very simple, and only requires 15 processors, as is illustrated in Figure 8-6. Three of these processors indicate a chord's type, where only three chord types (m7, 7, maj7) are required in the ii-V-I progression problem. The remaining 12 processors represent the chord's root pitch using pitch-class encoding. For example, Figure 8-6 demonstrates how one can represent the Dm7 tetrachord by only activating two units: the unit that represents that the chord is a minor seventh and the unit that indicates that the chord's root is the pitch-class D.

8.3.5 Implications

The sections above have discussed four different methods for encoding stimuli (and responses) for the ii-V-I progression problem. With these possible encodings of the ii-V-I progression described, we can now investigate the effect of problem encoding on network learning. Does problem representation affect network complexity? Does problem encoding alter the amount of training required for a network to solve the ii-V-I progression problem?

8.4 Complexity, Encoding, and Training Time

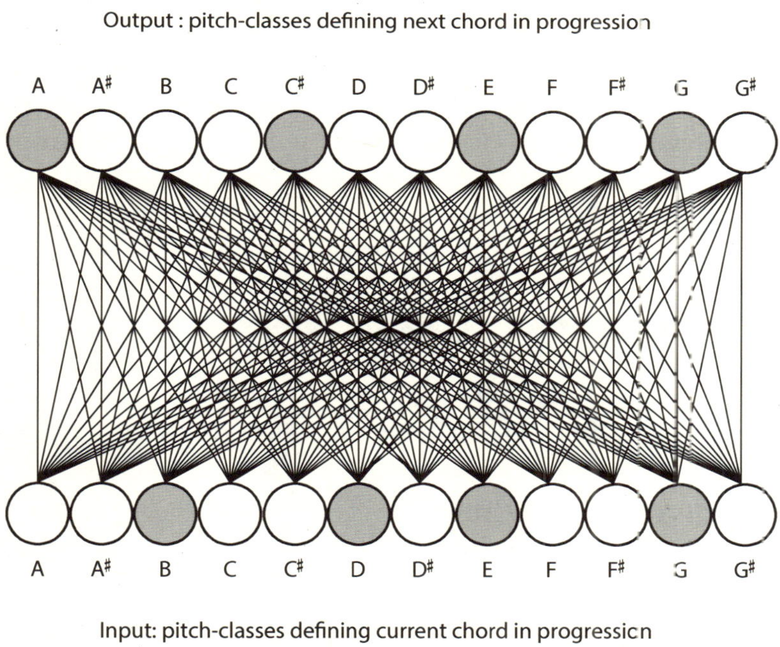

Figure 8-7 A perceptron trained on the ii-V-I progression task.

8.4.1 Task

All of the networks described in this section of this chapter learn the ii-V-I progression problem using one of the encodings discussed in Section 8.4. The question of interest concerns the effect of the various encodings: Does one encoding require a more complex network, or a greater amount of training, than another?

8.4.2 Network Architecture

In order to explore the effect of different encodings of the ii-V-I problem, pilot studies determine the simplest network capable of solving the problem for any of the encodings. Many of the musical networks reported in earlier chapters used value units. One reason for this is that value unit networks generally are easier to interpret. However, pilot studies indicate that networks of value units are challenged by the pitch-class encoding of the ii-V-I problem. A value unit perceptron could not learn this version of the problem; a multilayer network perceptron that had seven hidden value units was required. All three other encodings are learned by a value unit perceptron. However, all four encodings of the ii-V-I problem are also learned by perceptrons whose output units are integration devices that use the sigmoid-shaped logistic activation function. For the remainder of Section 8.4 we will consider how perceptrons that use integration devices as output units fare with the different encodings of the ii-V-I problem. Figure 8-7 illustrates one such network.

8.4.3 Training

All of the networks described below are trained with a gradient descent rule using the Rosenblatt program (Dawson, 2005). Each network is a perceptron whose output units use the logistic activation function. The only difference between a perceptron trained on one encoding of the ii-V-I progression problem and a perceptron trained on another encoding of this problem is the number of input and output units in the network. Perceptrons trained on a pitch-class encoding required 12 input and 12 output units. Perceptrons trained on a pitch encoding of non-inverted chords required 29 input and 29 output units. Perceptrons trained on a pitch encoding of inverted chords required 24 input and 24 output units. Perceptrons trained on a lead sheet encoding of the problem required 15 input and 15 output units.

When I train any perceptron on the ii-V-I progression problem, the learning rate is 0.50, and connection weights are randomly initialized to values in the range from −0.1 to 0.1. All output unit biases (θ) are set to zero, and do not change during learning. A network learns until it converges on a solution to the problem, where (as in previous chapters) convergence is defined as generating a hit for every output unit on every training pattern. As will be seen below, an integration device perceptron can learn any encoding of the ii-V-I problem very quickly.

8.4.4 Effect of Encoding

An integration device perceptron can learn each of the four encodings of the ii-V-I progression problem. As a result, none of the encodings makes this problem so complex that a more complicated network is required to solve it. However, there are definite effects of encoding on the amount of training that is required before a network converges on a solution. To explore these effects, one can train 20 different "subjects"—different networks—on each encoding of the ii-V-I progression problem. In this small experiment, the independent variable is the encoding of the problem and the dependent variable is the number of epochs of training required before the network converges. Table 8-2 provides the average number of epochs required (with standard deviations) for each of the four encoding conditions in this experiment.

Table 8-2 The mean number of sweeps required for a network to converge (with standard deviations) for perceptrons trained using four different encodings of the ii-V-I progression problem.

	Type of input encoding			
	Pitch-class	**Pitch (no inversion)**	**Pitch (inversion)**	**Lead sheet**
Mean	726.75	929.55	256.20	118.05
SD	1.02	2.14	2.02	0.22

Welch two-sample t-tests reveal that the difference between any pair of means in Table 8-2 is significant with $p < 0.001$. The smallest value of t is for the comparison between the means for the pitch encoding of inverted chords and the lead sheet encoding ($t = 304.63$, $df = 19.47$). The largest value of t is for the comparison between the pitch encoding of non-inverted chords and the lead sheet encoding ($t = 1687.28$, $df = 19.42$). In short, choice of encoding has a significant effect on the amount of training required for networks to converge. Pitch encoding of non-inverted chords results in the slowest learning; this suggests that pitch encoding provides the most complicated representation of the problem. If this encoding changes to pitch-class encoding, then a significant speeding up of learning occurs. One can produce another significant speeding up of learning by replacing pitch-class encoding with pitch encoding of inverted chords. Finally, another significant improvement in amount of training occurs when we use lead sheet encoding is used. Networks learn the lead sheet encoding of the problem with slightly less than one eighth of the amount of training required when pitch encoding of non-inverted chords is used.

8.5 Interpreting a Pitch-Class Perceptron

8.5.1 Integration Device Activity

Many of the networks that we have considered in earlier chapters used value units, which employ the Gaussian activation function. The simulations described in Section 8.4 revealed that perceptrons whose output units are integration devices are able to solve the ii-V-I progression problem. An integration device converts net input into activity using the sigmoid-shaped logistic function. Let us take a moment to consider the general properties of an integration device, and then use these properties to help interpret the internal structure of one of the perceptrons discussed in Section 8.4.

When the goal of a simulation is to interpret the internal structure of a network, value units have certain advantages. The Gaussian activation function responds to a particular subset of input properties. We have seen in earlier chapters that one can identify these properties by determining what input signals cancel out a value unit's μ, causing it to turn on.

In contrast, integration devices do not detect specific features that cause them to activate. Instead, they serve as devices that weigh evidence. Every signal coming into an integration device is either evidence in favour of turning on or evidence in favour of turning off. The activity of an integration device reflects the net effect of all of the accumulated evidence. It does not turn on when specific features are present. Instead, it turns on when enough positive evidence has accumulated.

There are two different perspectives for considering the meaning of an integration device's activity. The first is the digital perspective. We ordinarily train output integration devices either to turn on or to turn off. This requires the net input to an integration device either to be sufficiently high or to be sufficiently small. When the bias (θ) of an integration device is zero, as is the case for all of the Section 8.4 perceptrons, its net input must be 2.20 or higher in order for it to turn on (i.e., to generate activity of 0.90 or higher). If θ is zero, then turning an integration device off (i.e., to generate activity of 0.10 or less) requires a net input that is –2.20 or lower.

While training typically leads us to consider integration devices as being digital (i.e., as turning either on or off), we are not limited to the digital perspective. The continuous value of an integration device's activity ranges between zero and one, and is highly informative. In particular, an integration device's activity can be interpreted as a conditional probability (Dawson & Dupuis, 2012; Dawson et al., 2009; Dawson & Gupta, 2017; McClelland, 1998). Thus a second, analogue, perspective on interpreting integration device activity involves viewing activity as representing probability.

For instance, when an output integration device generates activity of 0.70, this means that there is a 70% chance that it will be "rewarded" given the current set of cues that have been presented to the input units (Dawson et al., 2009; Dawson & Gupta, 2017). We will see below that interpreting integration device networks benefits from considering such networks from the digital viewpoint as well as from the probabilistic perspective.

8.5.2 Network Interpretation

Let us consider the perceptron that solves the ii-V-I progression problem when pitch-class encoding of inputs and outputs is employed (Figure 8-7). This network's knowledge of the ii-V-I progression is stored in its connection weights. Figure 8-8 illustrates the connection weights from the 12 input units to a single output unit; the one that represents the pitch-class A. This pattern of connection weights is present for each output unit. The only difference between output units is that the bars are systematically assigned different pitch-class labels. For instance, the connection weights for the A♯ output unit can be plotted using exactly the same graph as Figure 8-8, but with different labels. For the A♯ output unit, the leftmost bar is associated with G♯ (instead of A), and the remaining bars are labelled A, A♯, B, and so on up to G. This implies that if we can explain the Figure 8-8 weights, then a functionally equivalent explanation applies to each of the other 11 output units.

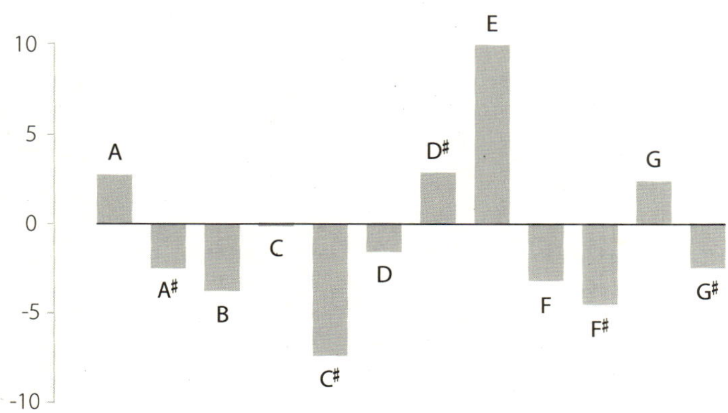

Figure 8-8 The connection weights from the 12 input units to the output unit representing the pitch-class A.

In previous chapters graphs of the connection weights that fed into value units revealed particular musical properties. The connection weights illustrated in Figure 8-8 feed into an integration device but do not reveal any obvious musical pattern. How then are these weights systematically used to turn the A output unit on when needed in the ii-V-I progression problem?

Let us first consider the A output unit from the digital perspective. What properties cause this unit to turn on? The ii-V-I progression problem consists of 24 different input tetrachords. Of these 24 patterns, eight cause the A unit to turn on, while the remaining 16 cause it to turn off.

Table 8-3 The eight patterns in the ii-V-I training set that cause the A output unit to activate when signals are sent through the weights illustrated in Figure 8-8.

Chord	Component pitch-classes	Net input
Am7	A, C, E, G	15.09
C7	A#, C, E, G	9.97
A7	A, C#, E, G	9.05
Em7	B, D, E, G	6.94
Cm7	A#, C, D#, G	2.82
F#m7	A, C#, E, F#	2.31
E7	B, D, E, G#	2.20
F7	A, C, D#, F	2.20

Note. Each row provides information about a particular chord. The first column names the chord, the next column provides the pitch-classes that make up the chord, and the final column provides the net input sent to the output unit when the chord is presented to the network.

Table 8-3 provides the features of the eight input patterns that turn the A output unit on. For each pattern, it provides the name of the input chord, the chord's four component pitch-classes, and the net input for the A output unit that is associated with each of these chords. The net input is simply the sum of the four weights associated with a chord's four input pitch-classes.

Not surprisingly, the net input column of Table 8-3 consists of values that are greater than or equal to 2.20, which is the net input required for an integration device to produce activity of at least 0.90 when $\theta = 0$. This suggests that each of these eight input chords are associated with four different connection weights whose sum is sufficient to activate the A output unit.

An inspection of the component pitch-classes in Table 8-3 provides an indication of why each of its rows is associated with high net inputs. For instance, six of these eight input patterns include the pitch-class E, which has the highest connection weight value by far in Figure 8-8. The two chords that do not include E (Cm7 and F7) include some combination of the pitch-classes A, D♯, and G, which are the other three connection weights with positive values.

In short, one account of the connection weights in Figure 8-8 is essentially combinatorial. The pitch-class input units are assigned these weights because 1) the values of the included weights produce high enough net input for the eight chords that turn the A input unit on, and 2) the values of the other weights produce low enough net input (–2.20 or lower) for the 16 chords that turn this output unit off.

Of the 12 connection weights depicted in Figure 8-8, eight are negative. Signals sent through a negative weight will decrease the A unit's activity. Is it the case that because of this some of the pitch-classes associated with negative weights are absent from Table 8-3? An inspection of the table indicates that this is not so. All 12 input pitch-classes appear at least once, although some (in particular A, C, E, and G) occur more frequently than the others. This suggests that the individual weights in Figure 8-8 might have more to do with the output unit from a probabilistic perspective. Perhaps strongly positive weights are not associated with pitch-classes that definitely turn the output unit on, but are instead associated with pitch-classes that are probably on when the output unit turns on.

To explore this possibility, let us consider the probability structure of the ii-V-I progression problem in the context of the A output unit. There are 24 different input patterns in this problem; eight of them cause the A output unit to turn on and 16 cause the A output unit to turn off.

Table 8-4 provides the number of times that each input pitch-class belongs to a pattern that turns the A output unit either on or off. For instance, the first row of Table 8-4 indicates that four of the input patterns that cause the A output unit to turn on include the input pitch-class A. Similarly, the first row indicates that four of the input patterns that cause the A output unit to turn off also include the input pitch-class A.

Table 8-4 also includes a "Conditional Probability" column. This column indicates the probability that a particular input pitch-class unit is on if one knows that the input chord turns the A output unit on. Thus, in the first row of the table, the conditional probability is P (Output A = 1|Input A = 1). This value is equal to 0.5 for the input pitch-class A because only four of the eight input patterns that include this pitch-class are associated with the A output unit turning on. Similarly, the conditional probability for input pitch-class A♯ is equal to 0.25. This is because only two of the eight input patterns that include this pitch-class are associated with the A output unit turning on.

Armed with a contingency table like Table 8-4, one would typically perform additional computations using Bayes' theorem to determine the conditional probability that the A output unit is on when a particular input pitch-class is used (e.g., we would compute P (Output A = 1|Input A = 1)). However, for this particular contingency table this probability is identical to the probability reported in the "Conditional Probability" column of the table.

With this knowledge of the probabilities of the ii-V-I problem, we can now explore network structure from a probabilistic perspective. Table 8-4 includes the weight of the connection between each input pitch-class unit and the A output unit. An inspection of these weights indicates that they seem related to the conditional probabilities in the table. That is, negative weights tend to be associated with lower probabilities, while positive weights tend to be associated with higher probabilities. The correlation between Table 8-4's "Conditional Probability" column and its column of connection weights is very high (r = 0.820).

As noted earlier, the "Conditional Probability" column in Table 8-4 can be interpreted as providing the probability that the A output unit is on given that a particular pitch-class input unit is turned on. The logistic activation function can also be used to estimate this probability. This is accomplished by sending a signal from only one input unit at a time into the output unit and examining the output unit's response. (This is equivalent to computing the logistic function with the input unit's weight as input, assuming that $\theta = 0$). The last column in Table 8-4 provides these probability estimates, which are even more strongly correlated with the "Conditional Probability" column than are the connection weights (r = 0.898). Clearly, the connection weights from Figure 8-8 encode the probability structure of the ii-V-I progression problem. In general, the more likely it is for an input unit to be involved in turning on an output unit, the larger will be the connection weight between the two.

However, while this story accounts for most of the network's structure, it is incomplete. This is why the correlations reported above are not perfect. Connection weights also have values that permit the network to deal with special cases.

Table 8-4 The probability structure of the ii-V-I progression problem in the context of the A output unit whose connection weights were presented in Figure 8-8.

Input pitch-class	Output A: On	Output A: Off	Conditional probability	Weight	Logistic of weight
A	4	4	0.5	2.81	0.94
A#	2	6	0.25	−2.31	0.09
B	2	6	0.25	−3.86	0.02
C	4	4	0.5	−0.15	0.46
C#	2	6	0.25	−6.19	0.00
D	2	6	0.25	−1.64	0.16
D#	2	6	0.25	2.81	0.94
E	6	2	0.75	9.97	1.00
F	1	7	0.125	−3.27	0.04
F#	1	7	0.125	−4.28	0.01
G	5	3	0.625	2.47	0.92
G#	1	7	0.125	−2.27	0.09

Note. Each row provides the structure related to a single input unit. The first column names the input unit. The second column indicates the number of patterns for which the input unit is on and the output unit for A is on, while the third column indicates the number of patterns for which the input unit is on and the output unit for A is off. The fourth column converts the preceding two columns into a conditional probability. The fifth column provides the weight between the input unit and the A output unit, while the sixth column converts that weight into activity using the logistic activation function.

For instance, consider the input unit that represents D♯. This input unit has a healthy positive connection to the output unit for A, even though D♯ has a low probability of turning A on (see Table 8-4). Given this low probability, why is this connection weight so strong?

The reason is the role of D♯ in the context of the three other pitch-classes that are also present to form a tetrachord. D♯ is present in six different chords that do not turn the A output unit on. For these six chords, D♯'s healthy positive weight is not an issue. This is because the other three pitch-classes in each chord have strong negative weights (producing, on average, a net input of −7.82). This is more than enough to cancel out the positive signal of D♯ and turn the A output off. For the two chords in which D♯ is present and the A output turns on, the other three pitch-classes are not positive enough (average net input = –0.30). The positive weight of D♯ is required to turn the A output unit on in these cases. D♯ has been assigned its strong weight to deal with these two cases.

In other words, the perceptron has learned weights that reflect the overall probability structure of the ii-V-I problem (a structure that simply considers the relationship between each input unit and an output unit without considering other input units). However, it then adjusts these weights so that the network generates the correct response in particular contexts (i.e., particular combinations of input signals) that would lead to incorrect responses if probability structure were the only consideration.

8.5.3 The Tonal Hierarchy

The tonal hierarchy is a key finding from Carol Krumhansl's research on music cognition (Krumhansl, 1990a), and played an important role in some of the key-finding perceptrons discussed earlier in Chapter 5. The tonal hierarchy reflects the differing importance of the various pitch-classes in the context of a particular musical key. For example, in the key of A major the tonic (the pitch-class A) receives the highest rating. The next highest ratings are given to the third or fifth positions of A major (the pitch-classes C♯ and E). The next lowest ratings are provided to the remaining four pitch-classes of the scale (for A major these are the pitch-classes B, D, F♯, and G♯).

The weights illustrated in Figure 8-8 also reflect the differing importance of various pitch-classes in a different context: in this figure, the context is the likelihood of activating the A output unit. This suggests that Figure 8-8 could be interpreted in a fashion similar to Krumhansl's (1990a) tonal hierarchy. What is the relationship between the tonal hierarchy and the weights illustrated in Figure 8-8? To answer this question, the correlation between the Figure 8-8 weights and the tonal hierarchy for each major key was computed. The results are presented in Table 8-5.

Table 8-5 reveals a definite relationship between Krumhansl's (1990a) tonal hierarchy and the Figure 8-8 weights. In particular, there is a very high correlation between the weights and the tonal hierarchy for the key of E major, which is a perfect fifth away from A. A smaller, but still healthy, correlation exists between the weights and the tonal hierarchy for the key of A major. Recall that the weights illustrated in Figure 8-8 are found for all of the output units in the Figure 8-7 perceptron, but are associated with different input units. As a result, the pattern of correlations reported in Table 8-5 is found for the connection weights that feed into the other

11 output units. The general finding for each output unit is a very high correlation with the tonal hierarchy a perfect fifth away from the output unit's pitch-class, and a high correlation with the tonal hierarchy that matches the output unit's pitch-class.

Table 8-5 The correlations between each of Krumhansl's tonal hierarchies for major keys and the connection weights illustrated in Figure 8-8.

Major key of tonal hierarchy	Correlation between tonal hierarchy and Output A weights
A	0.41
A#	−0.32
B	0.29
C	−0.09
C#	−0.32
D	0.06
D#	−0.10
E	0.73
F	−0.46
F#	−0.29
G	0.16
G#	−0.06

On the one hand, the relationship discovered between connection weights and the tonal hierarchy is surprising. The perceptron learns a problem that is quite different from the typical probe tone method. It is surprising, though satisfying, to see strong similarities to tonal hierarchies emerge from this network's internal structure.

On the other hand, the particular relationships revealed in Table 8-5 make perfect sense in the context of the probabilities provided in Table 8-4. For instance, the pitch-class that is most likely to be involved in turning the A output unit on is E. The probability relationships of Table 8-4 emerge quite naturally from the tonal hierarchy, given that the chords in the ii-V-I progression problem are all defined in particular major keys. Furthermore, within each major key all three chords (the minor seventh, the dominant seventh, and the major seventh) include pitch-classes that are a perfect fifth apart. Nevertheless, while the connection weights in Figure

8-8 strongly relate to the tonal hierarchy for E major, they are not perfectly correlated. How might differences between the two be reflected in musical structure?

Tonal hierarchies have been used to explore spatial relationships between different musical keys. One computes the similarity between two different keys by calculating the correlation between their respective tonal hierarchies. One then uses MDS to produce a map in which similar keys are closer to one another than are dissimilar keys (Krumhansl, 1990a; Krumhansl & Kessler, 1982). Krumhansl and Kessler found that this analysis arranges major keys in a map according to the circle of perfect fifths.

A similar analysis can be performed on the connection weights of the ii-V-I perceptron by comparing the similarities of the connection weights that feed into different output units. The correlations between each possible pair of sets of 12 connection weights are computed, and MDS is performed on this similarity data. This analysis arranges output units in a map; output units that have similar connection weight structures will be located close to one another. Figure 8-9 presents the two-dimensional solution when MDS is performed on connection weight similarities. This solution accounts for 28.92% of the variance in the original distances, which is a statistically significant fit ($F = 26.034$, $df = 1, 64$, $p < 0.001$). Unlike the tonal hierarchy analyses (Krumhansl & Kessler, 1982), this MDS solution does not arrange output units according to the circle of perfect fifths. Instead, two other organizational principles emerge. First, output units that represent pitch-classes a tritone away (e.g., B and F, A♯, and E) fall at nearly the same location in the map. Second, these pairs of tritones are near other pairs that are a semitone away. For instance, the nearest neighbours to the F-B pair are the F♯-C pair and the A♯-E pair, each of which contains a pitch-class that is either a semitone higher or lower than either F or B.

The two-dimensional MDS solution in Figure 8-9 reveals some intriguing regularities. However, an analysis that takes out more than two dimensions provides a better fit to the data. Figure 8-10 illustrates the first three dimensions of a five-dimensional solution for the output unit similarities. This solution accounts for 87.90% of the variance in the original distance data, which is a statistically significant fit ($F = 464.97$, $df = 1, 64$, $p < 0.001$).

Figure 8-10 indicates that the third dimension of this solution pulls the tritone pairs vertically apart from one another in the space. However, it is still clear that in this higher-dimensional solution pitch-classes a tritone apart are still located near one another in the space. The arrangement of points in Figure 8-10 corresponds quite nicely to the solution plotted in Figure 8-9. In fact, Figure 8-9 depicts what would be seen if one looked down on Figure 8-10 from above.

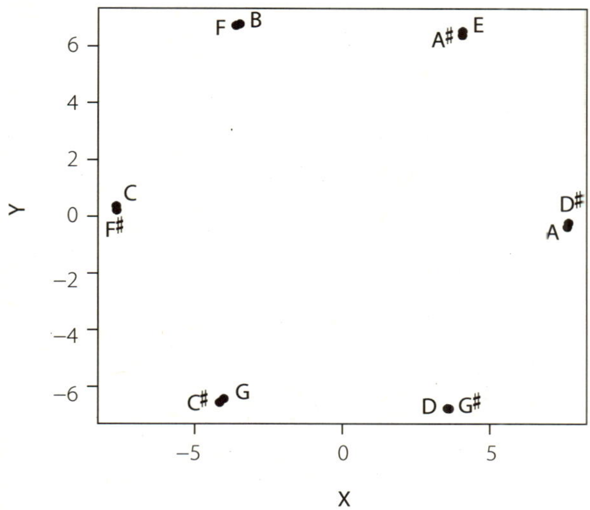

Figure 8-9 The two-dimensional MDS solution from the analysis of the similarities between output unit weights.

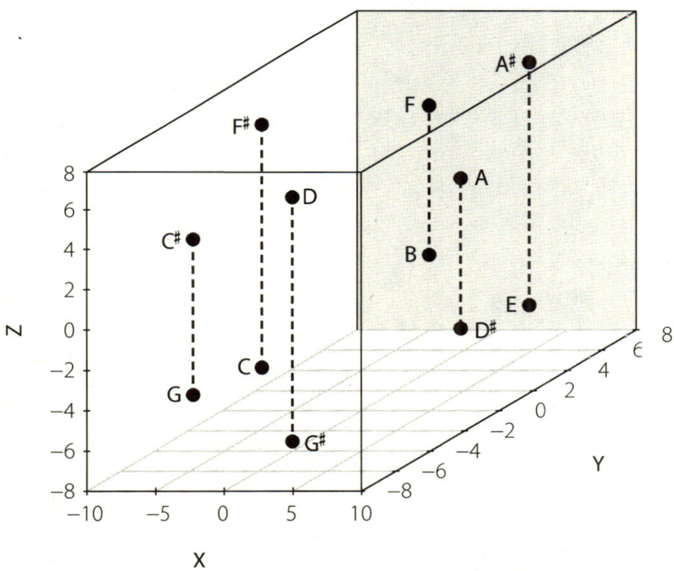

Figure 8-10 The first three dimensions of a five-dimensional MDS solution for the analysis of output unit weight similarities.

Why might tritone structure emerge from the ii-V-I progression problem? One answer to this question may be that the ii-V-I progression requires dominant seventh chords; these chords in turn make jazz's tritone substitution possible. Jazz uses chord substitutions to provide musical variety to the changes. In chord substitutions, one replaces a chord in a progression with another, musically related, chord. In tritone substitution, a dominant seventh chord in one key is replaced with the dominant seventh chord from a key that is a tritone away. For example, the ii-V-I progression in the key of A major uses the E7 chord. Under tritone substitution, it is replaced with A♯7.

Tritone substitution is possible because dominant seventh chords a tritone apart contain the same tritone. That is, they include the same two pitch-classes that are a tritone apart. This makes the two chords harmonically similar and permits them to be substituted for one another (Tymoczko, 2008). This harmonic similarity might also be the source for the tritone regularities in Figures 8-9 and 8-10. Even though the ii-V-I progression problem is not defined using tritone substitution, all of the possible dominant seventh chords are used. Their harmonic similarities—or the possibility of tritone substitution—are reflected in the structure of the two MDS solutions.

This suggests that it might be interesting to explore elaborations of the ii-V-I progression problem. For instance, one could define a version that explicitly defines tritone substitution in the training set. Would such a network exhibit similar structure to the one that we have been analyzing? Another kind of elaboration of the ii-V-I is a chord progression known as the Coltrane changes. Though related to the ii-V-I, the Coltrane changes are notoriously difficult to play. Can a network learn the Coltrane changes, using the various encodings introduced earlier in the chapter? If so, does the increased complexity of the Coltrane changes require us to use a more complicated network?

8.6 The Coltrane Changes

8.6.1 Extending the ii-V-I

The ii-V-I progression plays a dominant role in jazz. Shanahan and Broze compiled a corpus of 1200 jazz standards from published lead sheets (Broze & Shanahan, 2013; Shanahan & Broze, 2012). They analyzed this corpus to identify the five most common three-chord progressions in the lead sheets. The ii-V-I was by far the most prevalent, accounting for over 42% of the 7366 instances of these sequences.

Consider one jazz standard that employs the ii-V-I, "Tune Up." It famously appeared on the album *Blue Haze* recorded by Miles Davis for the Prestige label in sessions that took place in 1953 and 1954. Davis is credited as being the composer of "Tune Up" on this album, but it was actually composed by Eddie "Cleanhead" Vinson. Vinson was a prominent blues singer, saxophonist, and bandleader (Nisenson, 2000). Vinson, Davis, and "Tune Up" are all linked to another seminal jazz figure, saxophonist and composer John Coltrane. Coltrane was a member of Vinson's band in 1948. Coltrane also belonged to the Miles Davis Quintet between the years 1955 and 1957, as well as between 1958 and 1960 (Porter, 1998; Thomas, 1975). While in the Davis quintet, Coltrane was involved in performances and recordings of "Tune Up" (DeVito & Porter, 2008).

While jazz is founded on core harmonic patterns such as basic chord progressions, it always pushes this core in new directions. Coltrane was a master of this pursuit. His instructor at the Granoff School in Philadelphia, Dennis Sandole, reported that he and Coltrane investigated many advanced harmonic concepts that served as the foundation for Coltrane's landmark compositions (Demsey, 1991). Some of Coltrane's harmonic experiments led to a jazz progression now known as the Coltrane changes. This progression was unveiled in his influential 1960 album *Giant Steps*, where it appears in two famous pieces, "Giant Steps" and "Countdown." The Coltrane changes also appear in several other pieces that Coltrane composed around this time (Demsey, 1991). The title "Countdown" pays homage to the Vinson–Davis classic "Tune Up"; Demsey shows that the harmonic structure of "Countdown" is systematically linked to that of "Tune Up." Indeed, the structure of Coltrane's changes can be explained as a particular elaboration of the ii-V-I progression (Demsey, 1991). The Coltrane changes add four new chords to the ii-V-I progression. Two of these added chords serve as lead-ins to the V chord in the ii-V-I, while the other two added chords are lead-ins to the I chord in the ii-V-I.

Importantly, the relationship between the roots of the V chord and its two lead-ins is a musical interval of a major third; the same relationship holds among the roots of the I chord and its two lead-ins (Demsey, 1991). The circle of perfect fifths and the four circles of major thirds can be used to create a map of the Coltrane changes chord roots for any key. Figure 8-11 provides this map. The inner circle of pitch-classes in this figure is organized around the circle of perfect fifths. Each of these pitch-classes is then attached to a circle of major thirds, which forms the outer ring of pitch-classes. These circles of major thirds provide the roots of the lead-in chords.

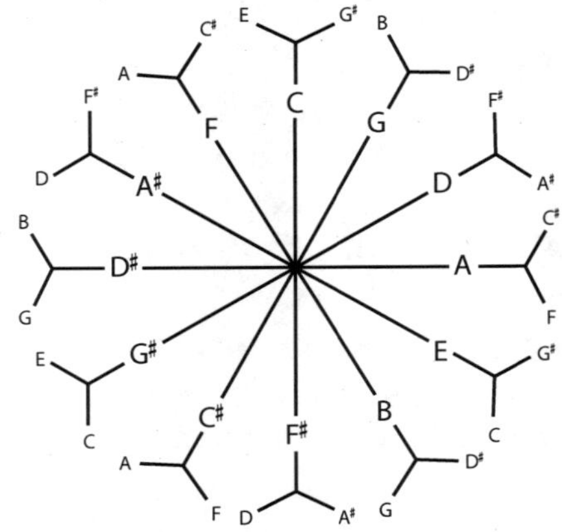

Figure 8-11 A map of the Coltrane changes' chord roots created by combining the circle of perfect fifths (the inner ring of pitch-classes) with the circles of major thirds.

To illustrate the use of this map, consider the Coltrane changes for the key of C major. The ii-V-I progression for C major is Dmin7-G7-Cmaj7. The Coltrane changes elaborate this sequence by adding two lead-in chords for G7 and two lead-in chords for Cmaj7. Figure 8-12 presents the part of Figure 8-11 used for the key of C major, naming each chord and providing the order in which they are played. From Figure 8-12 one can see that the Coltrane changes for C major are Dmin7-D♯7-G♯maj7-B7-Emaj7-G7-Cmaj7. Note that this progression begins with the first chord of the ii-V-I, and it ends with the ii-V-I's final two chords.

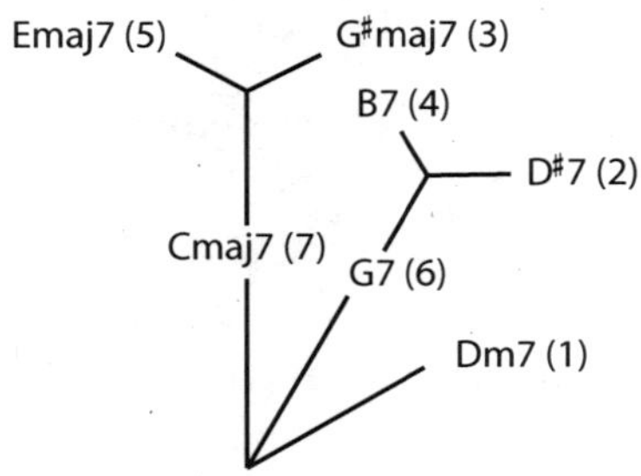

Figure 8-12 The portion of the Figure 8-11 map that provides the Coltrane changes for the key of C major.

Figure 8-11 can be used to determine the seven chords of the Coltrane changes for any major key; one simply finds the key's root in the inner circle and builds a version of Figure 8-12 up from this root. Table 8-6 provides the complete set of Coltrane changes, each of which was determined by applying this method for each of the 12 major keys. Note that the ii, V, and I columns of Table 8-6 also provide the ii-V-I progression for these major keys.

Table 8-6 The Coltrane changes for each major musical key.

Major key	Chord						
	ii	1st lead-in for V	1st lead-in for I	2nd lead-in for V	2nd lead-in for I	V	I
A	Bm7	C7	Fmaj7	G#7	C#maj7	E7	Amaj7
A#	Cm7	C#7	F#maj7	A7	Dmaj7	F7	A#maj7
B	C#m7	D7	Gmaj7	A#7	D#maj7	F#7	Bmaj7
C	Dm7	D#7	G#maj7	B7	Emaj7	G7	Cmaj7
C#	D#m7	E7	Amaj7	C7	Fmaj7	G#7	C#maj7
D	Em7	F7	A#maj7	C#7	F#maj7	A7	Dmaj7
D#	Fm7	F#7	Bmaj7	D7	Gmaj7	A#7	D#maj7
E	F#m7	G7	Cmaj7	D#7	G#maj7	B7	Emaj7
F	Gm7	G#7	C#maj7	E7	Amaj7	C7	Fmaj7
F#	G#m7	A7	Dmaj7	F7	A#maj7	C#7	F#maj7
G	Am7	A#7	D#maj7	F#7	Bmaj7	D7	Gmaj7
G#	A#m7	B7	Emaj7	G7	Cmaj7	D#7	G#maj7

Note. Each row provides the sequence of chords that define this progression for a particular musical key; the column labels indicate the role of each chord.

8.6.2 The Coltrane Changes Problem

With Table 8-6 in hand, I can now define a new chord progression problem using the Coltrane changes. As was the case with the earlier simulations using the ii-V-I chords, I present one chord from the progression to a network, and it generates the next chord in the progression. This requires six input/output pairs for any major key.

The problem involves learning these chord pairings for each major key, producing a training set composed of 72 different patterns (six for each major key).

8.6.3 Encodings

As was the case in our study of the ii-V-I progression problem, we can explore a number of different encodings for the Coltrane changes problem: pitch-class, pitch without inversions, pitch with inversions, and lead sheet. Interestingly, the structure of the Coltrane changes produces some challenges for some encodings that were not present when the simpler ii-V-I problem was encoded. For instance, consider some additional design decisions that are required when pitch encoding without inversions is used. Unlike the ii-V-I progression, the Coltrane changes can cover a very wide piano keyboard. The number of input units that are required to represent this width depends on where various chords are started on the keyboard. This must be considered because the same chord can occur on different places in the keyboard depending on the musical key in which the changes are being defined. This issue is addressed by stipulating that every tetrachord in the training set has at least one pitch present in the first octave of a piano keyboard. That is, one can shift every chord down the keyboard so that its lowest note is in the range from A3 (the A below middle C) to G♯4 (the G♯ above middle C). Using this encoding, the non-inverted forms of all of the chords used in the Coltrane changes require 23 input units. The lowest input unit corresponds to A3, and the highest input unit corresponds to G5.

A different complication arises when encoding the Coltrane changes using pitch representations of chord inversions. There are two basic issues: First, what chord inversions should be used, and second, where should they be placed on the represented keyboard? Using voice leading as a guide to choice of inversions, I selected the set of chord forms that are summarized in Table 8-7. I then used two different methods to place the chords on the "keyboard," producing two different versions of training sets. In the first, I defined the chord patterns for the lowest major key on the keyboard. Then I shifted this set of chord patterns up the keyboard to define the Coltrane changes for all of the other major keys. This requires 23 input units to represent any input/output pairing. In this representation the lowest pitch represented is A3 (the A below middle C) and the highest is G♯4 (the G♯ above middle C). This choice of encoding means, unlike the encoding of non-inverted chords, that the same chord is represented in different octaves. However, all of these different representations of the same chord involve representing the chord in a different form (i.e., a different inversion). In the second method, I used the same method used for the pitch encoding of non-inverted chords: every chord was shifted downward so that at least one note belongs to the lowest octave of the input pitches.

Table 8-7 The various chord forms used to achieve efficient voice leading for the Coltrane Changes.

Chord	Chord form
ii	First inversion
First lead-in for V	Root position
First lead-in for I	Second inversion
Second lead-in for V	First inversion
Second lead-in for I	Third inversion
V	Third inversion
I	First inversion

8.7 Learning the Coltrane Changes

8.7.1 Relative Complexity

Section 8.4 reported the results of training artificial neural networks on the ii-V-I progression problem using a variety of encodings. Although we discovered that choice of encoding affected the amount of training required for a network to converge, the major finding of that section was that a very simple network—a perceptron with integration devices in its output layer—could learn any version of the ii-V-I progression problem. We have seen that the Coltrane changes elaborate the ii-V-I progression, and that the Coltrane changes are more difficult for musicians to perform or to improvise over. Are they more difficult for networks to learn?

The following sections report the results of training networks on various encodings of the Coltrane changes, and show that choice of encoding can have an important effect on network complexity. In general, though, all of these simulations point to one general conclusion: the Coltrane changes are indeed more complicated than the ii-V-I progression. This is because an integration device perceptron was never capable of learning a solution to the Coltrane changes, regardless of the choice of encoding. The Coltrane changes require using a more complex network.

8.7.2 Pitch-Class Encoding

In all of the simulations reported in this section, I attempt to discover the simplest network capable of learning the Coltrane changes. With pitch-class encoding, a multilayer perceptron is required. This perceptron uses value units for its output units, and requires nine hidden value units in order to converge. During training, its connection weights are randomly initialized in the range from –0.1 to 0.1, and μs are initialized to zero. The learning rate is 0.01, and each μ is modified by training. The order of pattern presentation is randomized for each epoch of training.

A simulation experiment was conducted in which 25 different networks were trained until convergence was achieved; convergence involved generating a "hit" for every output unit and every training pattern. In this simulation experiment convergence was achieved after a mean of 5299.12 epochs of training (*SD* = 2774.75). The fastest convergence was obtained after only 1771 epochs of training, while the slowest convergence required 13,044 epochs.

8.7.3 Pitch Encoding without Inversions

When I use pitch encoding to encode the Coltrane changes chords in their root form, a much simpler network is able to learn the problem: a perceptron that uses value units in the output layer. A simulation experiment was conducted in which 25 different networks of this type were trained to convergence. The network was initialized and trained in an identical fashion to the multilayer perceptron described in Section 8.7.2, with the exception that a learning rate of 0.1 was used. On average, convergence was obtained after 301.08 epochs of training (*SD* = 61.81). The fastest convergence was obtained after 168 epochs, while the slowest convergence required 380 sweeps of training.

While a perceptron learns this version of the Coltrane changes, the fact that this perceptron uses value units instead of integration devices indicates that this encoding of the Coltrane changes is still more complicated than the ii-V-I progression problem. This is because a value unit makes a more complicated carving of a pattern space, making two parallel cuts through it instead of just one (Dawson, 2004, 2005, 2008).

8.7.4 Pitch Encoding with Inversions

As noted in Section 8.6.4, I studied two different versions of the inverted Coltrane changes. Let us first consider the version in which we did not shift chords to all have at least one note in the first octave of the pitch representation. This version of the problem is very difficult in comparison to the non-inverted version of the Coltrane changes. A multilayer perceptron with 11 hidden units is required; all of the output units and all of the hidden units of this network are value units. The structure of this network is initialized in the same way as those discussed above, and it is trained with a learning rate of 0.01. A simulation experiment in which 25 different networks were trained revealed that convergence was obtained after an average of 5237.68 epochs of training (SD = 1487.00). This amount of training is not significantly different from the amount required by the networks trained on the pitch-class representation of the problem (t = 0.0976, df = 36.735, p = 0.9228). It is, however, significantly greater than the amount of training required for the perceptron given the non-inverted encoding (t = −16.5849, df = 24.083, p = 0.001).

The version of the Coltrane changes in which inverted chords were shifted downward to start in the first octave of inputs also requires a multilayer perceptron that is built with value units and contains 11 hidden units. A simulation experiment revealed that this network took much longer to converge than did the other version of the inverted Coltrane changes. On average, 8299.92 epochs of training were required to reach convergence (SD = 3446.14). This amount of training is significantly greater than the amount required by the other version of the inverted chords (t = −4.0794, df = 32.638, p = 0.001).

It appears, then, that using inverted chords makes learning the Coltrane changes a much harder task. First, a more complicated network is required. Second, more training is required. One likely reason for this result is that inverting the chords requires networks to learn more causal links between chord forms than is required when chords are not inverted. In addition, shifting the chords down to all start from the same octave makes the task even more difficult. This is likely because this shift disrupts causal relations between chords even further.

8.7.5 Lead Sheet Encoding

The lead sheet encoding of the Coltrane changes leads to the most efficient learning of this progression. As with the pitch representation of non-inverted chords, a value unit perceptron learns the lead sheet version of the problem. A simulation experiment in which 25 of these networks were trained revealed that on average convergence was achieved after 72.36 epochs of training (SD = 1.89). This is a significantly smaller amount of learning than is required by the perceptron presented the non-inverted chords (t = 18.4924, df = 24.045, p = 0.001).

In summary, all of the results described above support two general conclusions. First, the Coltrane changes are more difficult than the ii-V-I progression because, regardless of encoding, they cannot be learned by an integration device perceptron.

Second, choice of encoding of the Coltrane changes has a marked effect on network learning. This choice determines both network complexity and the amount of training required to achieve convergence.

8.8 Interpreting a Coltrane Perceptron

8.8.1 Coltrane Causality

How does an artificial neural network represent its knowledge of the Coltrane changes? To answer this question let us interpret the internal structure of a value unit perceptron that learns this progression using lead sheet encoding. Before examining the network, it will be useful to understand the causal structure that links chords in the Coltrane changes, which we can infer from examining Table 8-6.

First, because lead sheet encoding separates chord types from chord roots, let us consider the causal links between chord types in the Coltrane changes. There are only three relationships between chord types. First, a minor seventh chord always causes the next chord to be a dominant seventh. Second, a dominant seventh chord always causes the next chord to be a major seventh. Third, a major seventh chord always causes the next chord to be a dominant seventh. Let us next consider causal links between chord roots. These causal links are mediated by chord type, but we can ignore this context for the time being.

First, a chord root can cause the next chord root to be a minor second or one semitone higher. For instance, the first row of Table 8-6 shows that the first transition for the Coltrane changes in the key of A major is from a chord with the root of B to a chord with the root of A.

Second, a chord root can cause the next chord root to be a perfect fourth or five semitones higher. For example, Table 8-6 shows that this happens three times in the key of A major: C causes F, G♯ causes C♯, and E causes A.

Third, a chord root can cause the next chord root to be a minor third or three semitones higher. For instance, Table 8-6 shows that this happens twice in the key of A major, because F causes G♯, and C♯ causes E.

When causal links involving chord types and causal links involving chord roots are considered in combination, very systematic causal relations emerge in the Coltrane changes. First, causal links between specific chords are unique. For instance, C7 always precedes Fmaj7. An examination of Table 8-6 indicates that any chord of interest only precedes one chord.

Second, this property means that there are chains of chord sequences that are repeated in different keys of the Coltrane changes. One example chain is C7 – Fmaj7

– C♯7 – G♯7 – C♯maj7. This sequence of chords appears in the first row of Table 8-6, where C7 is the first lead-in for the V chord in the key of A major. The same sequence is also found in the fifth row of Table 8-6, where C7 is the second lead-in for the V chord in the key of C♯ major.

In short, the Coltrane changes can be described as a set of systematic and unique causal links in which the occurrence of one chord in a network's input units necessarily causes the occurrence of a specific chord in the output units. In order to "know" the Coltrane changes, a network must adjust its connection weights in such a way as to realize these causal links. In the next section, we discover how a value unit perceptron accomplishes this.

8.8.2 Network Structure

The network to interpret is a value unit perceptron trained on the lead sheet encoding of the Coltrane changes. This particular perceptron is initialized in the same fashion as was described earlier in this chapter, and was trained with a learning rate of 0.1. However, in order to facilitate network interpretation I hold the μ of each output unit to zero throughout training. As a result, for an output unit to turn on, its net input needs to be near zero in value. The network converges after 93 epochs of training. Table 8-8 presents the resulting connection weights.

8.8.3 Network Causality

Table 8-8 The connection weights for a perceptron that has learned the Coltrane changes in lead sheet notation.

Input unit	Output unit													
	m7	**D7**	**maj7**	**A**	**A#**	**B**	**C**	**C#**	**D**	**D#**	**E**	**F**	**F#**	**G**	**G#**
μ	0.00	0.00	0.00	0.23	−0.79	0.22	0.24	−0.25	−0.79	0.79	−0.25	0.23	0.24	−0.24	−0.22
m7	−0.98	0.07	1.04	2.10	1.45	2.09	2.12	−2.09	−1.78	1.78	−2.11	2.12	2.13	−2.11	−2.09
D7	−1.06	−1.23	−0.19	1.11	−1.76	1.10	1.12	−1.11	1.42	−1.43	−1.12	1.12	1.13	−1.12	−1.10
maj7	−1.03	0.07	1.09	0.78	0.76	0.78	−1.12	−0.74	1.78	0.26	−0.74	0.77	0.77	−0.77	−0.78
A	−0.36	−0.07	0.19	0.77	−0.26	−0.17	0.75	1.11	−0.26	−1.77	−0.74	0.77	0.77	−0.77	−0.78
A#	−0.24	−0.07	0.19	0.78	−0.26	0.78	−0.20	−0.74	−1.41	0.27	2.11	0.77	0.77	−0.77	−0.78
B	−0.24	−0.07	0.19	0.77	−0.26	0.78	0.75	0.19	−0.26	1.42	−0.75	−2.11	0.77	−0.77	−0.78

Input unit	Output unit														
	m7	D7	maj7	A	A#	B	C	C#	D	D#	E	F	F#	G	G#
C	−0.25	−0.07	0.19	0.78	−0.25	0.79	0.75	−0.74	0.76	0.26	1.11	0.77	−2.12	−0.77	−0.78
C#	−0.27	−0.07	0.19	0.78	−0.26	0.79	0.75	−0.74	−0.26	−0.75	−0.74	−1.11	0.77	2.11	−0.78
D	−0.23	−0.07	0.19	0.77	−0.26	0.78	0.75	−0.74	−0.26	0.27	0.19	0.77	−1.12	−0.77	2.09
D#	−0.25	−0.07	0.19	−2.10	−0.25	0.78	0.75	−0.74	−0.26	0.27	−0.74	−0.18	0.77	1.11	−0.78
E	−0.22	−0.07	0.19	0.78	−1.44	0.78	0.75	−0.74	−0.26	0.27	−0.75	0.77	−0.19	−0.77	1.10
F	−0.32	−0.07	0.19	−1.10	−0.26	−2.09	0.75	−0.74	−0.26	0.26	−0.75	0.77	0.77	0.19	−0.78
F#	−0.23	−0.07	0.19	0.77	1.76	0.78	−2.11	−0.74	−0.26	0.27	−0.75	0.77	0.77	−0.76	0.17
G	−0.26	−0.07	0.19	−0.18	−0.25	−1.09	0.75	2.09	−0.26	0.26	−0.74	0.77	0.77	−0.76	−0.78
G#	−0.26	−0.07	0.19	0.23	−0.79	0.22	0.24	−0.25	−0.79	0.79	−0.25	0.23	0.24	−0.24	−0.22

Note. Each row corresponds to an input source (μ or an input unit) and each column corresponds to an output unit. Unique connection weights from input to output are highlighted in grey.

How do the connection weights in Table 8-8 represent the Coltrane changes? They do so by instantiating all of the specific causal relationships that were introduced in the previous section. First, consider the causal relationships that link an input chord type to an output chord type. These relationships are enforced by the weights presented in the first three columns of Table 8-8. In the Coltrane changes, an input minor seventh chord causes an output dominant seventh chord. Two aspects of the connection weights bring this condition to life. First, the connection between the input unit for m7 and the output unit for D7 has a weight of 0.07. Second, the connection between every chord root input unit and the output unit for D7 has a weight of −0.07. As a result, when the m7 unit is activated at the same time as a chord root input unit, the signals from the two input units to the D7 output unit cancel out to zero, turning this input unit on.

A dominant seventh chord can also be activated by a major seventh chord. The network accomplishes this in exactly the same way as was described in the preceding paragraph: note that the connection weight from the maj7 input unit to the D7 output unit is also equal to 0.07. Its signal, when combined with the signal from a chord root input unit, produces a net input of zero that again turns the D7 output unit on.

The network uses the same connection weight logic to activate the maj7 output unit when the D7 input unit is activated. The connection weight from this particular input unit to this particular output unit is −0.19. The connection weight between any chord root input unit and the maj7 output unit is 0.19. As a result, the D7 input unit will combine with any input chord root unit to produce a net input of zero that activates the maj7 output unit.

Importantly, the network also assigns connection weights to the first three columns of Table 8-8 in such a way that output units do not turn on when they are supposed to be off. For example, an input chord never turns on the m7 output unit, because the minor seventh is the chord that starts the Coltrane changes in a given key. Note that no combination of input signals in the first column of Table 8-8 will produce a net input of zero. Similarly, the connection weights from the "wrong" chord types to either the D7 or the maj7 output units are extreme enough never to be cancelled by a signal coming from any input chord root unit.

Let us now turn to considering how the weights in Table 8-8 handle causal links involving chord roots. To do so let us consider the output unit for pitch-class A. The connection weights that feed into this output unit are presented in the fourth column of the table. This output unit is activated by three different causal links between chords: E7 – Amaj7, F♯maj7 – A7, and G♯m7 – A7.

The first clue as to how these causal links are instantiated by a perceptron comes from examining the connection weights in Table 8-8 from each of the chord root units to the A output unit. All but three of these weights are approximately 0.77. The only three exceptions involve the pitch-classes involved in the three causal relationships described above. The weight from E is −2.10, the weight from F♯ is −1.10, and the weight from G♯ is −0.18.

The second clue to chord root causality comes from the relationship between these three different weights to the weights between the A output unit and the three chord type input units. First, the weight from the D7 input unit to A is 2.10. This exactly cancels the signal coming from the E input unit. In other words, when D7 and E are both activated, the A output unit will turn on.

A similar relationship holds for the other two input chord types. The weight from the maj7 input unit to the A output unit is 1.10, which cancels out the signal from the F♯ input unit. As well, the weight from the m7 input unit to the A output unit is 0.23, which essentially cancels out the signal coming from the G♯ input unit.

In short, it appears that for the A output unit to turn on, the two input units must be activated at the same time. One is a chord type unit; the other is a unit representing the chord root. The pairing of chord type and chord root is exactly as required by the causal relationships involved in turning the output unit on.

An examination of the remaining columns in Table 8-8 reveals the same connection weight logic. All but three of the connection weights have the same value. The three that have unique values come from input chord roots involved in turning a particular output unit on. The unique values serve to cancel out the signal coming from a particular chord type unit. These unique values are highlighted in grey in the table. Note that this pattern of grey is very systematic, tracing out a pattern of three diagonals through these weights. This pattern reflects the systematic relationships between input and output chord roots when the musical intervals between these roots are considered (see Section 8.8.1).

8.9 Strange Circles and Coltrane Changes

8.9.1 Circles of Major Thirds

When I use lead sheet encoding for a chord progression problem, a chord's type and a chord's root are encoded separately. This provides a network with an opportunity to determine independently an output chord's type and root. The analysis of the Coltrane changes perceptron in Section 8.9 indicates that the network does not take this independent path. Instead, the network makes explicit the specific causal relationships between pairs of chords. For example, instead of asserting that 7 – maj7 and C – F, the network makes explicit the more specific relationship that C7 – Fmaj7. This claim is supported by the fact that turning on any chord type output unit, or turning on any chord root output unit, requires combining signals from both a chord type input unit and a chord root input unit.

This approach to solving the Coltrane changes points toward a new direction for representing the structure of this chord progression. Figure 8-11 combined the circle of perfect fifths with the four circles of major thirds to generate a map of the Coltrane changes in any key. An alternative approach to generating the Coltrane changes in a particular key is to use only the circles of major thirds.

Two aspects of the network interpretation inform this approach, which is described in detail below. First, a particular input chord (i.e., a combination of chord type and chord root) always causes a particular output chord. Second, a particular input chord never causes an output chord where the roots of the input and the output chords belong to the same circle of major thirds. For example, in Table 8-8, output chords with the root A are only caused by input chords with the roots E, F♯, or G♯ (the grey cells in the A column). Output chords with the root A are never caused by input chords with the roots C♯ or F, which belong to the same circle of major thirds as does A.

This suggests that the Coltrane changes is a sequence of chords in which one chord (associated with one circle of major thirds) can only cause a subsequent chord that is associated with a different circle of major thirds. Interestingly, this means that there is a very simple algorithm that uses three different circles of major thirds to generate the seven chords of the Coltrane changes in a particular major key (Figure 8-13).

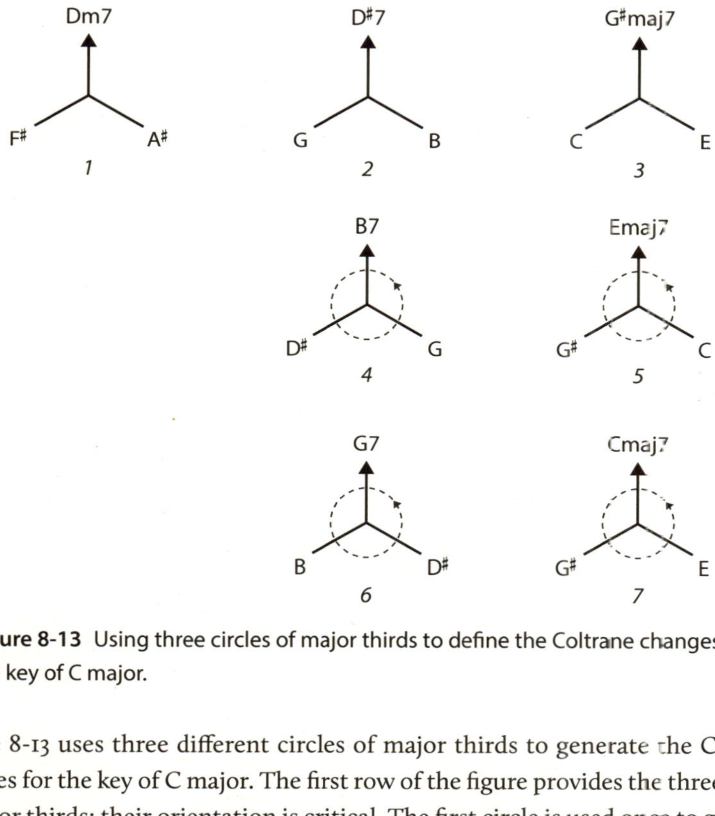

Figure 8-13 Using three circles of major thirds to define the Coltrane changes for the key of C major.

Figure 8-13 uses three different circles of major thirds to generate the Coltrane changes for the key of C major. The first row of the figure provides the three circles of major thirds; their orientation is critical. The first circle is used once to generate the minor seventh chord that begins the progression. The chord that is played is at the top of the circle, and is pointed to by an arrow. The second circle is used to generate the dominant seventh chord. It is first used to select the D♯7 chord in the top row of the figure, again a chord pointed to by an arrow. The third circle is used to generate the major seventh chord. It is first used to select the G♯maj7 chord in the top row of the figure.

Importantly, to generate the remaining chords in the progression, one first rotates each of the second two circles of major thirds counter-clockwise by 120°. This brings two new chords to the top of these circles, as illustrated in the second row of the figure. These chords are then played in succession; the second circle is used to generate the next dominant seventh chord (B7) and the third circle is used to generate the next major seventh chord (Emaj7). Then these two circles are rotated again counter-clockwise by 120°. This brings the final two chords to the top of these circles, as is shown in the third row of the figure.

The three circles of major thirds illustrated in Figure 8-13 can be used to generate the Coltrane changes for two other musical keys as well. If one takes each circle in the top row and rotates it counter-clockwise by 120 degrees, then the circles will generate the Coltrane changes for the key of G♯ major if the preceding algorithm is used. If the circles in the top row of Figure 8-13 are each rotated counter-clockwise by 240 degrees before starting, then they will produce the Coltrane changes for the key of E major.

With different combinations of circles of major thirds, one can produce the Coltrane changes for any major key. The specific combinations are provided in Figure 8-14. The first row of this figure provides three circles oriented to generate the Coltrane changes for the key of C major (as was illustrated in Figure 8-13); rotating each of these circles once clockwise or twice clockwise orients them for generating the progression for G♯ major and E major respectively. The second row is oriented to produce the progression for A♯ major, and can be rotated to accommodate the keys of F♯ major and D♯ major. The third row is oriented for B major, and can be rotated to produce the chord sequence for G major and D♯ major. The final row is oriented for C♯ major, and can be rotated to produce the Coltrane changes for A major and F major.

Figure 8-14 is interesting because it shows that each circle of major thirds appears three times; each time it appears it does so in a different column. As well, each circle of major thirds is missing from only one of the four rows in the figure. Figure 8-14 is also of interest because it points in certain directions related to jazz composing. First, because the Coltrane changes are comprised of seven chords, the first circle of thirds is only used to generate one chord. In other words, while the Coltrane changes define two lead-in chords to the V chord of the ii-V-I, as well as the I chord of this progression, they do not define lead-in chords for the ii chord. Figure 8-14 provides a strong motivation for elaborating the Coltrane changes by adding two new chords, both minor seventh chords that lead in to the ii. The chords in question are provided by the two unused pitch-classes in the first column of the figure. Second, the fact that each row of Figure 8-14 is missing a circle of major thirds leads one to consider adding it to provide up to three new chords for the progression. Some musical exploration is required to determine where in Figure 8-14 one would insert this new source of chords, as well as to determine the kind of chord to associate with this additional component.

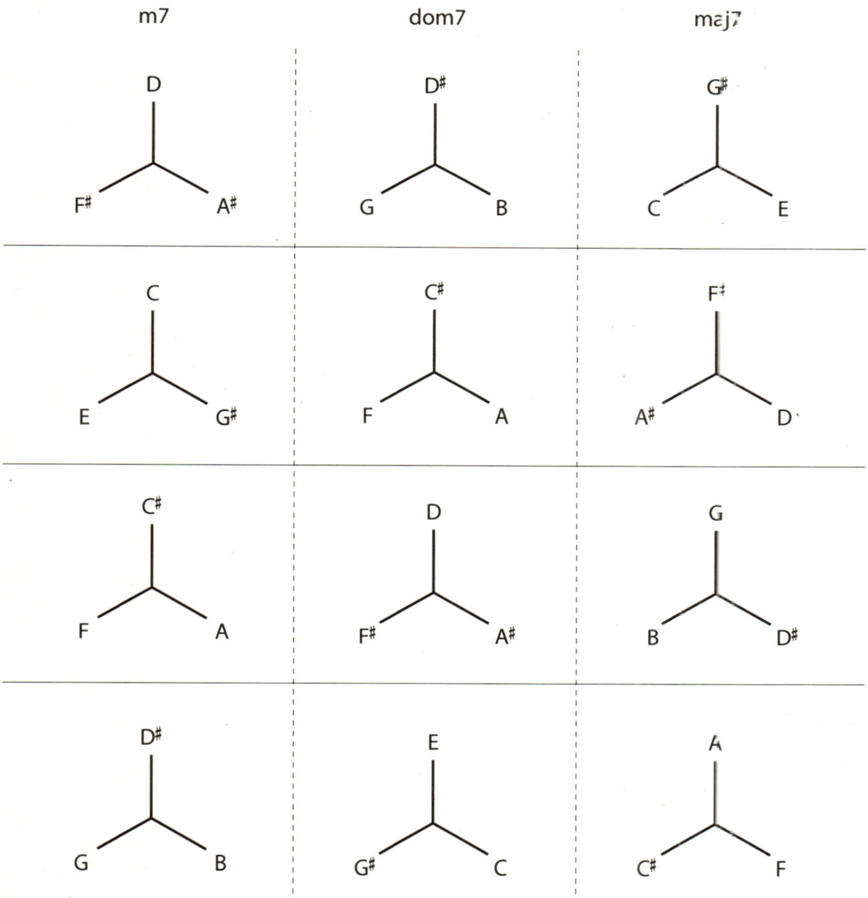

Figure 8-14 The circles of major thirds for generating the Coltrane changes in any key.

8.10 Summary and Implications

At the start of this book, I introduced artificial neural networks as artifacts primarily used for pattern classification. That is, they arrange input patterns as points in a space (either a pattern space or a hidden unit space), and output units carve this space into decision regions. If a pattern falls into one decision region, the network generates one kind of response (i.e., one kind of "pattern name"); if it falls into a different decision region, a different response is generated.

In earlier chapters, I have demonstrated that pattern classification is a general ability that can be applied very neatly to a variety of musical problems. For example, I have used it to identify scale tonics, scale modes, musical keys, and chord types. The current chapter has shown a further flexible use of pattern classification in which the response generated by a network to an input chord is a special name: the name of another chord. This permits a network to represent chord progressions in its internal structure. I demonstrated this ability by training networks on two different chord sequences, the ii-V-I progression and the Coltrane changes.

In addition to demonstrating this ability, this chapter also explored the importance of how one encodes network stimuli and responses. One of the main results obtained in the current chapter was that the choice of encoding had enormous impact on problem complexity. For the ii-V-I progression problem, I discovered that encoding did not affect network complexity: all versions of this problem could be learned by an integration device perceptron. However, the choice of encoding did affect the amount of training required for a network to discover a solution to this problem. For the Coltrane changes, choice of encoding not only affected learning speed but also determined network complexity. Some versions of this problem could be solved by a value unit perceptron, while others required multilayer networks of value units that included 11 hidden units.

While a main purpose of the current chapter was simply to illustrate the importance of encoding choices, it is important to keep in mind the implications of such choices. Obviously, problem difficulty is dictated by problem encoding. What encoding, then, should one choose for their networks? It might be very tempting to explore a variety of different and plausible encodings, and then to choose the one that generates the simplest networks. In some cases, this might very well be the appropriate strategy. However, other factors must also be considered when choosing an encoding. For example, perhaps the goal of a network is to provide insight into the formal regularities that govern a specific musical problem. In this case, the encoding that leads to the simplest network may not be the most appropriate, because the encoding may cause certain musical regularities to disappear.

We saw earlier in this chapter that one key element of the musical theory of chord progressions is voice leading. The lead sheet notation described in this chapter generates simple networks, but this encoding hides essential properties related to voice leading. Therefore, if one is interested in using networks to explore regularities of voice leading, then the encoding that leads to the simplest network may not be the most appropriate.

As another example, perhaps the goal of training a musical network is to discover representations that serve as the basis for musical cognition. In this case, we may not be searching for the encoding that produces the simplest networks. Instead, we might be searching for the encoding that generates the greatest similarity between various measures of network performance and structure and measures of performance of human listeners in a musical cognition experiment.

From the perspective of musical cognition, human listeners are "black boxes." This is because we cannot directly observe the internal structures and processes that mediate musical cognition. Instead, we can only infer these internal properties from observations of external behaviour. This process of inference is known as reverse engineering. By observing human responses to musical stimuli in a variety of clever experimental situations, we attempt to discover the structures, processes, or algorithms inside the black box.

Reverse engineering is hard enough because we cannot directly see inside the black box. A second issue that makes reverse engineering challenging is that each input/output or stimulus/response pairing that we can observe can be mediated by more than one process. There is a many-to-one mapping from possible structures, processes, or algorithms to input/output relations (Dawson, 2013). As a result, we might believe that one process is responsible for mediating observed behaviour, but a very different process might actually be responsible. Therefore, we require some special observations useful for validating one theory about what is inside the black box as opposed to another. Fortunately, black boxes will generate some observable behaviours that are side effects of the processes inside the black box. These side effects—called artifacts by Dawson (2013)—can provide critical information for theory validation (Pylyshyn, 1980, 1984).

For instance, one consequence of representing a problem in a particular format might be that some instances of the problem can be solved quickly, while other instances are more difficult to solve. In performing mental arithmetic, for example, one might expect that if numbers were represented mentally in columns then addition problems that require carrying digits from one column to another would take longer than problems that did not require this operation. One can collect relative complexity evidence (Pylyshyn, 1984) to investigate artifacts of this type. With relative complexity evidence, one varies the nature of problems presented to a system, and then explores the relationship between the properties of the problems and the time required to solve them.

A related type of data provides intermediate state evidence (Pylyshyn, 1984). This kind of evidence presumes that information processing inside the black box requires a number of different processing stages, and that each stage might represent intermediate results in a different format. To collect intermediate state evidence, one attempts to determine the number and nature of these intermediate results. For example, when researchers determined that items in short-term memory were confused with similar sounding items (Conrad, 1964) and not with items with similar meaning, this suggested that an intermediate memory store used an acoustic encoding (Waugh & Norman, 1965).

A particular type of data, called error evidence (Pylyshyn, 1984), is very well suited to determine intermediate states. When extra demands are placed on a system's resources, it may not function as designed, and its internal workings are likely to become more evident (Simon, 1969). This is not just because the overtaxed system makes errors in general, but because these errors are often systematic, and their systematicity reflects the underlying representation. For example, one study (Yaremchuk & Dawson, 2005) investigated a multilayer perceptron trained to identify tetrachord types. When some of its hidden units were removed, the network only made very specific errors: it failed to identify tetrachords as being major when, and only when, they were in their second inversion form. This suggested that the role of the missing hidden units was to permit the network to deal with this rather specialized type of input.

What is the relationship between relative complexity evidence, intermediate state evidence, error evidence, and choice of encoding? In many cases, researchers are specifically interested in using artificial neural networks to serve as models of human musical cognition (Griffith & Todd, 1999; Todd & Loy, 1991). In this case, establishing the validity of the model likely requires collecting all three types of evidence, not only from the human subjects but also from the neural network model. The hope would be to find a close relation between the evidence collected from the human subjects and the evidence collected from the neural network model. Importantly, this match is likely to be highly related to choice of encoding. In other words, a music cognition researcher may not be interested in seeking the encoding that leads to the simplest network, but instead in seeking the encoding that leads to the best match between subject and model.

9

Connectionist Reflections

9.1 A Less Romantic Connectionism

The Overture that began this book fancifully asked whether it is true that two systems that generate the same musical inputs and outputs must also share an underlying theory of music. Do the aliens inside the mother ship in *Close Encounters of the Third Kind* require a theory of Western music to jam with the human scientists below? Alternatively, is it possible that some alien musical theory can also produce identical musical patterns? The purpose of this book was to explore this issue, replacing fictional organisms from another world with agents that are more practical: artificial neural networks. I trained networks to perform a number of tasks that mapped musical stimuli to responses that are well defined in Western music theory. After successfully teaching these networks, I conducted in-depth analyses of their internal structure. In general, I discovered formalisms inside these networks that differed from those that are typical of Western music. The purpose of this chapter is to step back and consider the general results that have been detailed in the preceding chapters, and to discuss the implications of these results. However, before embarking on this summary, let us first consider the relationship between the connectionist research we have been considering and more typical studies that employ artificial neural networks.

9.1.1 A Romantic Revolution

Connectionist cognitive science erupted in the mid-1980s with the discovery of learning rules capable of training networks with a layer of hidden units (Ackley, Hinton, & Sejnowski, 1985; Rumelhart et al., 1986). Connectionism, as a reaction to classical cognitive science, has many parallels with the Romanticist reaction against the age of reason (Dawson, 2013). Two of these parallels are of particular interest to us as we reflect upon the results presented in this book.

First, as noted in Chapter 2, when connectionist cognitive science arose it explicitly abandoned theories that appealed to the rule-governed manipulation of symbols. When researchers introduced artificial neural networks by training them to perform prototypically classical tasks, such as changing the tense of verbs (Rumelhart & McClelland, 1986a) or solving logic problems (Bechtel & Abrahamsen, 2002), this was done to demonstrate that such tasks did not require using explicit rules and symbols. Connectionism is Romanticist in its abandonment of the formal or logical.

Second, connectionist cognitive science attacked the classical view for proposing theories that were neither neuronally inspired nor biological plausible. A central tenet of connectionism is that intelligence emerges from the unique and complex interactions among vast numbers of nonlinear neurons (Churchland, 1986; Churchland & Sejnowski, 1992; Searle, 1984). However, in appealing to biologically plausible information processing, connectionists moved toward theories that were nearly impossible to elaborate fully. When inspired by the brain, connectionists are swept away by the sublime, revealing their Romanticism.

Connectionism's rejection of the formal and its acceptance of the sublime are further cemented by the fact that connectionists rarely provide detailed interpretations of their networks' internal structures. Perhaps this is particularly true of the limited connectionist literature on music cognition. Many researchers believe that a key advantage of artificial neural networks is their ability to adapt to musical regularities that cannot be formalized (Bharucha, 1999; Rowe, 2001; Todd & Loy, 1991).

Recent developments in artificial neural network research take connectionist Romanticism even further. In the 1980s, the connectionist revolution began with networks that used only one or two layers of hidden units. Nowadays there is a growing sense that such networks are inadequate for adapting to complex, real world situations. There is an emerging interest in new set of techniques called deep learning that allows researchers to train networks with many layers of hidden units (Bengio, Courville, & Vincent, 2013; Hinton, 2007; Hinton, Osindero, & Teh, 2006; Hinton & Salakhutdinov, 2006; Larochelle, Mandel, Pascanu, & Bengio, 2012). However, the interpretation of deep networks is even more challenging than the interpretation of shallower networks (Erhan, Courville, & Bengio, 2010). Thus, as connectionism turns more and more to deep learning, it becomes even more Romanticist in nature.

9.1.2 Reducing Romanticism

The research presented in this book moves in the direction opposite to deep learning. I have trained very shallow networks—some having no hidden units at all—on very simple musical tasks, employing very straightforward representations. This approach reacts against connectionist Romanticism. I am not concerned with informal, sublime musical properties. Instead, this research project explores the ability of shallow networks to capture formal musical regularities. Furthermore, I am particularly interested in whether networks can reveal new formal properties that are not typical of modern music theory. This book has explored the viability of an anti-Romanticist use of artificial neural networks.

This exploration was largely inspired by one of the seminal publications in musical cognition, Carol Krumhansl's *Cognitive Foundations of Musical Pitch* (Krumhansl, 1990a). Krumhansl used perceptual experiments to explore her subjects' responses to musical stimuli. Recognizing the enormous breadth of possible musical stimuli, not to mention the range of possible musical responses and the variety of potential musical expertise, Krumhansl shaped and streamlined her research by making a number of important design decisions. She developed paradigms that permitted responses to musical stimuli to be obtained easily. She focused on responses to a limited but well-established domain, musical pitch, typically creating her musical stimuli from a simple and tractable set of building blocks: the 12 pitch-classes of Western tonal music. Krumhansl's research led to a number of fundamental insights into music and musical cognition. These insights were possible because of her design decisions.

Before conducting our simulations, I made a number of design decisions that were similar in spirit to those made by Krumhansl. First, I employed supervised learning rules for our artificial neural networks. This is because I wanted to know exactly what networks learned to do to facilitate the interpretation of their internal structure. In supervised learning, networks converge to a solution only after they generate the desired (and known) response to each input pattern.

Second, all of our simulations trained networks to generate simple, well-established regularities of Western tonal music. Again, with a solid formal understanding of what networks learned to do I was hoping to be better positioned to interpret their internal structure. These tasks included identifying the tonic and the root of musical scales, key-finding, identifying the types of different triads and tetrachords, and generating chord progressions.

Third, most of our simulations used very simple input and output representations. Many of the networks described in earlier chapters used pitch-class encoding, which meant that very simple networks—networks that only had 12 input units—could learn a musical task of interest. Our choice of this representation paralleled Krumhansl's (1990a) assumption of octave equivalence in her cognitive studies of musical pitch.

Fourth, all of our simulations sought the simplest networks capable of converging upon a solution to the task. Simple networks are easier to interpret. By seeking the simplest networks, I was able to make some surprising discoveries. For instance, identifying a scale's tonic, performing key-finding, or generating the next chord in a jazz progression, can all be accomplished by simple perceptrons that do not require any hidden units.

Fifth, the point of each of our simulations was not simply to create a network to solve a particular musical task. Instead, the point was to interpret the internal structure of a trained network in order to determine the solution to the task that it had discovered.

All of the research presented in this book explores a basic question: can shallow networks trained on simple musical problems reveal anything novel about the structure of music? In general, I can say that the answer to this question is a resounding *yes*. Let us now reflect on the various results reported in the preceding chapters.

9.2 Synthetic Psychology of Music

9.2.1 Synthetic Psychology

Cognitive science is mostly conducted using an analytic strategy called reverse engineering (Cummins, 1983; Dawson, 2013). In reverse engineering, data is collected from a behaving system, and then some type of model—a model of data, a mathematical model, or a computer simulation—is fit to the data (Dawson, 2004). The purpose of the model is to provide a concise account of the regularities in the data. Reverse engineers collect data first, and fit the model to the data second. However, an alternative research approach called forward engineering is also available to cognitive scientists (Dawson, 2004, 2013; Dawson et al., 2010a). Forward engineering is also known as synthetic psychology (Braitenberg, 1984). In forward engineering, a model is first put together from a basic set of interesting components. Then the behaviour of the model is observed in various situations. In forward engineering, the model does not describe data that has already been collected, but instead is the source of the data. Synthetic psychologists build their models first, and only collect data second—from the model they have constructed.

Artificial neural networks have been proposed as an ideal medium in which synthetic psychology can be conducted (Dawson, 2004). When used in forward engineering, the basic components of artificial neural networks (activation function, learning rule, etc.) are the basic building blocks used to construct the model. Because the network is trained on what to do, but is not told specifically how to accomplish this task, networks can be a source of surprising algorithms or representations that can be used to propose novel theories in cognitive science. However, for connectionist synthetic psychology to succeed, it requires substantial reverse engineering after synthesizing a network. The surprising revelations produced by connectionist forward engineering reveal themselves only after understanding the internal structure of a trained network. The insights that networks provide are not found in their behaviour (the input/output mappings they produce), but instead exist in the regularities that networks have discovered to mediate their behaviour.

This book serves as a case study in connectionist synthetic psychology. First, I forward engineer networks to solve basic musical problems. Second, I reverse engineer the networks to determine how they solve these problems, and to relate these methods to traditional music theory. The success of this approach is measured by the nature and number of surprising representations that I discover in the trained networks. The following sections summarize these major discoveries. The surprises that I revealed concern the complexity of networks trained on various tasks, the suitability of value units for musical tasks, the importance of the tritone, the use of strange circles, and the properties of distributed representations of music.

9.2.2 Network Complexity

One of the mysteries confronting a connectionist forward engineer at the start of a project concerns network complexity. What kind of network is required to learn a task? Is a multilayer perceptron necessary? If so, how many hidden units does it require? When one begins a simulation project, one typically has certain expectations about network complexity. Surprises often occur when these expectations are shown to be false.

One example of this occurred in the initial simulations involving scale tonics and scale modes (Chapters 3 and 4). Conventional music theory defines a very specific pattern of musical intervals between adjacent notes in both a major and a harmonic minor scale. It does not provide a similar general definition for the tonic of such a scale. Because of this, I expected that classifying a scale's mode would be an easier problem than classifying its tonic. However, our simulations demonstrated that this expectation was false. Scale tonic identification is an easier problem that can be solved by a perceptron. Classifying scale modes is more complex, and requires the use of a multilayer perceptron. A second example of this sort of surprise was provided in our exploration of the ii-V-I progression problem. The expectation was that the pitch-class encoding of this problem would require a multilayer perceptron.

It was astonishing when this version of the problem could be learned by a perceptron that used integration devices for output units.

In general, it is interesting and perhaps somewhat surprising that small networks can solve all the various musical problems that I have considered, even when I use pitch-class encoding. The most complicated networks that were encountered (the more abstract encodings of the Coltrane changes) required between nine and 11 hidden units. The remainder of the multilayer perceptrons reported in the book required far fewer hidden units.

9.2.3 The Value of Value Units

One reason for the relative simplicity of the networks that I have reported is that many of them use value units. I opted for value units because they offer many advantages over more traditional architectures when network interpretation is involved (Berkeley et al., 1995; Dawson, 1998, 2004, 2005, 2013; Dawson & Boechler, 2007; Dawson et al., 2005; Dawson et al., 2000a; Dawson et al., 2000b; Nickerson, Bloomfield, Dawson, Charrier, & Sturdy, 2007). However, the simplicity of most of the networks suggests that the value unit architecture is particularly well suited for the synthetic psychology of music. Perhaps this is because the activation function of a value unit is tuned so that the unit only turns on to a very narrow range of net inputs (Dawson & Schopflocher, 1992). As a result, in most of my simulations value units learned to respond to a very small number of musical patterns. Clearly, this tuned sensitivity of the architecture facilitated network interpretation. However, it also permitted very simple networks to learn the problems that I defined. It appears that underlying almost all of the tonal tasks that I studied is a definite relationship between certain musical properties and a desired musical judgment.

This is not to say that the particular musical properties exploited by my networks are traditional or unsurprising. Most of the network interpretations revealed a very different logic underlying some aspect of Western tonality. These novel formalisms are summarized below.

9.2.4 The Prominent Tritone

There appears to be a bias against the tritone in Western music. A long history of studying the consonance of the various musical intervals has indicated that the tritone is one of the most dissonant. For many years, researchers have studied the perceptual properties of the various musical intervals (Bidelman & Krishnan, 2009; Guernsey, 1928; Helmholtz & Ellis, 1863/1954; Krumhansl, 1990a; Malmberg, 1918; McDermott & Hauser, 2004; McLachlan, Marco, Light, & Wilson, 2013; Plantinga &

Trehub, 2014; Plomp & Levelt, 1965; Seashore, 1938/1967). Perhaps it is because of its dissonance that the tritone is one of the least frequently appearing intervals, both in Western music and in the music of other cultures (Vos & Troost, 1989). Indeed, the rarity of the tritone is one of the reasons that its presence may be important for key-finding (Browne, 1981; Butler, 1989).

The networks that I have explored do not appear to share this bias against the tritone. Starting with the analysis of the scale mode network, one of the surprises that emerged from interpreting musical networks is their strong preference for the tritone. Repeatedly I found that networks took advantage of the fact that two pitch-classes were a tritone apart to structure their responses to musical stimuli. Table 9-1 tabulates the many examples of tritone usage that we have encountered.

Table 9-1 Examples from previous chapters of identifying tritone relationships in a variety of network interpretations.

Task	Regularity	Depiction
Detecting scale mode	Tritone balance in both hidden units	Figures 4-3 and 4-4
Detecting scale mode	Grouping of minor scales with identical balanced tritones in hidden unit space	Figure 4-2 and table 4-1
Triad classification	Tritone balance in hidden unit weights	Figures 6-4 and 6-5
Classifying added note tetrachords	Tritone equivalence in hidden units weights	Figure 6-24
Classifying extended tetrachords	Tritone equivalence in hidden unit weights	Figures 7-8 and 7-10
Classifying extended tetrachords	Tritone balance in hidden unit weights	Figures 7-11 and 7-12
ii-V-I progression problem	Tritone organization of weight space in MDS solution	Figures 8-9 and 8-10

I repeatedly encountered two general types of tritone exploitation. The first is tritone balance, in which two pitch-classes a tritone apart are assigned connection weights that are equal in magnitude but opposite in sign. As a result, when input units representing each of these pitch-classes are simultaneously active their signals cancel each other out, typically increasing processor activity when value units are part of a network's architecture.

The second is tritone equivalence, in which two pitch-classes a tritone apart are assigned identical connection weights. As a result, in terms of network processing both of these pitch-classes are functionally identical. Tritone equivalence frequently appears in networks trained on harmonic tasks like chord classification.

Tritone balance and tritone equivalence are characteristics of connection weights between input units and other processors. It is not surprising that when these tritone regularities are seen, I also find tritone organization in analyses that are more abstract. For instance, plots of points in hidden unit spaces, or from various MDS analyses of weights or unit activities, organize themselves so that points related by a tritone are close together in the space.

All of these results lead naturally to a key question: Why do so many musical networks exploit the tritone? One possibility is that the tritone, which is a musical distance of six semitones, divides the octave exactly into two. Perhaps the networks discover that many musical tasks can be solved by identifying the same regularities in each half of the octave.

9.2.5 Strange Circles

The many examples of tritone equivalence provided in Table 9-1 illustrate another surprising property revealed in many network interpretations: the use of strange circles. A strange circle involves an equivalence class of pitch-classes that are related to each other by a specific musical interval, such as the six pitch-classes that define a circle of major seconds.

Network usage makes these circles "strange" because in a variety of circumstances different pitch-classes that belong to the same musical circle are assigned the identical connection weight. As a result, to the network these different pitch-classes are functionally identical. Table 9-1's listing of tritone equivalences picks out the occasions in which the strange circles are based on the tritone. Table 9-2 provides the instances of strange circles based on other musical intervals that have been encountered in our network interpretations.

One interesting property revealed by Table 9-2 is that the use of strange circles based on intervals other than the tritone only seems to emerge for tasks involving harmonic stimuli. While strange circles of tritones appear in other tasks, the additional circles only appear when networks learn to classify triads or tetrachords.

Table 9-2 Examples from previous chapters that discovered use of strange circles in different network interpretations.

Task	Regularity	Depiction
Triad classification	Circles of minor thirds	Figures 6-3 and 6-27
Triad classification	Circles of major thirds	Figures 6-4 and 6-5
Triad classification	Circles of major seconds	Figure 6-6
Classifying extended tetrachords	Circles of major seconds	Figure 7-5
Classifying extended tetrachords	Circles of minor thirds	Figures 7-7 and 7-8
Classifying extended tetrachords	Circles of major thirds	Figure 7-9

9.2.6 Distributed Representations

One of the major contributions of connectionism to the study of cognition has been the proposal for alternative forms of mental representation. Perhaps the most important of these connectionist contributions has been the notion of coarse coding or of distributed representation (Hinton et al., 1986; Pollack, 1990; Thrun, 1995). Although distributed representations are very difficult technically to define (Van Gelder, 1991), intuitively they involve simultaneous activities in a number of different hidden units; these activities are combined to produce a correct response. Coarse codes are interesting because each of the active components seem as though they have poor sensitivity to properties related to making correct judgments. In a distributed representation, each component is an inaccurate detector, but the combination of these poor components leads to high accuracy.

Distributed representations have been repeatedly encountered when interpreting musical networks. In two notable instances networks appear to solve musical problems by seeking intersections between groups of possibilities picked out by various inaccurate hidden unit detectors.

One example of this type of processing occurred when a multilayer perceptron was trained to perform key-finding (Section 5.4.2). Plots of each hidden unit's responses to the various keys revealed that each was a very inaccurate detector of musical key (Figure 5-5). However, if one sought the intersection of the sets of keys picked out by each hidden unit's activity, then the correct musical key could be isolated.

A second example of coarse coding was revealed in the examination of extended tetrachord classification. One reason for using the value unit architecture in many of our simulations is that such units often produce bands of activity where each level of activity captures different subsets of input patterns (Berkeley et al., 1995; Dawson, 2004; Dawson & Boechler, 2007; Dawson & Piercey, 2001). This in turn can facilitate network interpretation. The hidden units in the extended tetrachord network demonstrated distinct banding (Section 7.3). Each of these bands picked out different subsets of extended chord types, again demonstrating the inaccuracy of detection by each individual hidden unit. However, if one determined the intersection of the different sets of chords picked out by each hidden unit's band, then the correct type of tetrachord was the result.

The two instances provided above are the most prototypical examples of coarse coding that were revealed in my simulation studies. However, a more liberal notion of distributed coding permits us to claim that several other examples of this type of representation were discovered. For instance, on several occasions I described input stimuli as points in a hidden unit space, where the activity of each hidden unit to a pattern provides the coordinates of its location. Output units generate correct responses to problems by carving the hidden unit space into decision regions. Importantly, a hidden unit space is a distributed representation because the location of any point in this space depends upon considering the activity in each hidden unit simultaneously.

From this perspective, hidden units are not required to create distributed representations; such representations are in perceptrons as well. For example, activities passing through every connection weight in a scale tonic perceptron must be considered in order to determine the tonic of an input scale. Furthermore, the set of weights in that perceptron combines the properties of major scales and harmonic minor scales into a single (distributed) representation. Similarly, signals sent through all the weights of an ii-V-I perceptron provide a distributed representation of conditional probabilities related to specific output pitch-classes.

9.2.7 Summary

The results reviewed in this section indicate that my musical networks have yielded a number of interesting and surprising regularities. Even though these networks learned tasks that can be defined using traditional music theory, they have discovered non-traditional means for mediating their input/output mappings.

Why might these results be of interest? The final sections of this chapter consider the implications of these results for two different domains: music and musical cognition.

9.3 Musical Implications

Section 9.2 provided a general overview of my simulation results. Simple artificial neural networks can easily be trained to perform musical tasks that are based on Western tonality. In addition, the internal structure of these networks can be interpreted; these interpretations reveal formal musical regularities. Many of these regularities provide interesting departures from traditional music theory. Ignoring musical cognition for the time being, what are the implications of such results for the study of music in general?

9.3.1 Levels of Investigation

Our consideration of these implications will be aided by recognizing that cognitive science investigates phenomena at different levels of analysis, each of which requires a special vocabulary to capture particular kinds of regularities (Dawson, 1998, 2013; Pylyshyn, 1984). Following the lead of computational vision pioneer David Marr (Marr, 1982), the most abstract level of analysis is the computational level. At this level, researchers investigate what kind of information processing problem is being solved by a system of interest. The computational level of analysis typically uses formal methods that provide proofs that answer this question.

The second level of investigation is the algorithmic level. At the algorithmic level, researchers are typically concerned with determining the particular information processes involved in solving an information-processing problem. That is, what algorithm or procedure is being used to solve an information-processing problem identified at the computational level of analysis? Experimental paradigms, like those developed by cognitive psychologists, typically provide the methods required to perform an investigation at the algorithmic level.

Marr's third level of investigation is the implementational level. For Marr, this was the level where the methods of neuroscience explained how the information processes identified at the algorithmic level are brought into being by the brain. In modern cognitive science, it is useful to consider two separate questions related to implementation. The first is the architectural level of investigation. At this level, one determines the most basic information processes that are wired into the brain (e.g., primitive symbols and primitive rules). Once this has been determined, an implementational analysis à la Marr can be conducted to explain how the architecture is built into the brain. As far as the relationship between our musical networks and music is concerned, the computational and algorithmic levels of analysis are highly relevant. Let us consider our network contributions in the context of these two levels.

9.3.2 Theory Informs Algorithms

In connectionist cognitive science, the computational level of analysis is concerned with defining the input/output mapping performed by an artificial neural network. At this most abstract level, a network is a device that computes a mathematical function that converts input information into output information. The computational level of analysis defines the function being computed.

From this perspective, music theory itself defines and provides the input/output functions that networks were trained to generate in the preceding chapters. Identifying a scale's tonic or mode, or classifying triads or tetrachords into chord types, all involve functions whose formal structures are completely defined by music theory.

The tasks described in the preceding paragraph are formal but are not typically expressed mathematically. Fortunately, the formal apparatus of modern music theory permits mathematical definitions of these input/output mappings. Mathematical set theory was applied to music beginning in the 1960s in order to describe atonal music (Babbitt, 1960, 1961; Forte, 1973, 1985; Lewin, 2007; Straus, 1991). While aimed at atonal music, the properties that set theory formalizes can also be used to describe regularities in tonal music.

Indeed, it seems natural to consider that the function of many of my trained networks is to perform set theory operations. One of the basic operations in musical set theory is to express a musical pattern in normal order. A network that is presented a scale in pitch-class representation, and then delivers its tonic, can be thought of as a device that renders the stimulus into a set of elements in normal order, and then returns the first element in that set. A network presented a scale in the same format, but which delivers the scale's mode, can be considered a device for assigning something akin to a Forte number to the input pattern.

9.3.3 Algorithms Inform Theory

Section 9.3.2 suggests that music theory in general, and musical set theory in particular, provides an appropriate formalism for a computational account of my musical networks. At the very least, music theory defines the training sets that I created for our networks. While the computational level defines what input/output mapping is being computed, analysis at the algorithmic level—network interpretation—reveals how this mapping is mediated by a network's internal components (Dawson, 1998, 2004, 2013). We have seen that the analysis of musical networks can reveal formal properties that are quite different from those used to define their training set. For example, one of the main results of the algorithmic analyses summarized in Section 9.2 was the discovery that networks often use strange circles to solve harmonic problems. In a strange circle, different pitch-classes that belong to a circle defined by a particular musical interval (e.g., circles of major thirds, major seconds, or tritones) are all treated as being the same pitch.

Musical set theory uses one strange circle—the circle of octaves—when it makes the assumption of octave equivalence. This assumption limits the basic elements of set theory to the 12 different pitch-classes. The strange circles discovered in the musical networks point to a radically different set of basic elements. For instance, circles of major seconds reduce one to considering only two different kinds of elements, pitch-classes that belong to one circle or pitch-classes that belong to the other. Similarly, a set theory based on circles of major thirds would consider only four different kinds of elements, because pitch-classes can belong to only one of four different circles.

If one were to develop a musical formalism based on one or more of the strange circles, then it seems obvious that it would be quite different from musical set theory. However, it might be both interesting and viable. After all, the networks appear to use such a theory to classify types of chords. For another example, consider a second major finding reported in Section 9.2, the discovery of tritone balance in a variety of networks. Unlike tritone equivalence, tritone balance means that the signal generated by a unit representing one pitch-class is cancelled by the signal generated by the unit representing the pitch-class a tritone away.

Tritone balance has some interesting implications for musical set theory. After assuming octave equivalence, music set theorists then order the elements that define a musical stimulus in a particular way. For instance, Forte numbering assigns the pitch-class C the value 0, the pitch-class C♯ the value 1, and so on. This means that in music set theory pitch-classes are organized around a circle of minor seconds (Figure 6-9).

In the circle of minor seconds, pitch-classes that are a tritone apart are opposite one another across the diameter of the circle. Tritone balance occurs when there is a special relationship between these opposite pitch-classes. To make this relationship explicit in music set theory one might first adopt a different numbering system. For instance, if C is assigned the number x, then F♯—a tritone away from C—could be assigned the number –x. Additional operations on sets, involving sums of these numbers, would then have to be invented to take advantage of whatever tritone balance might offer.

The previous examples have shown how certain properties discovered from network interpretations might inform musical set theory. Importantly, networks offer other information pertinent to the computational consideration of music. For example, there is a long history of generating maps that represent the similarity between notes or scales in terms of the distances between points (Krumhansl, 2005; Schoenberg, 1969; Tymoczko, 2012). My musical networks provide a variety of new properties for generating maps that have different arrangements than those mentioned above. For instance, instead of measuring scale similarity in terms of shared pitch-classes, one can measure scale similarity in terms of connection weights. Similarly, at many points in preceding chapters we considered hidden unit spaces. These spaces are alternative maps of musical stimuli in which the coordinates of each point in the map are provided by hidden unit activities.

The point of considering different sorts of network-derived maps is that in many instances they might arrange musical stimuli in a fashion that is quite different from that found in other musical maps. By exploring these differences—by considering why certain musical entities are close to one another and why others are not—it is possible to develop alternative musical theories.

Musical networks can also provide evidence related to other computational issues. For example, one general type of question that often arises in computational analyses concerns the complexity of one situation in comparison to another. Network training provides one approach to answering such questions. In our simulations, because of our interest in network interpretation, we sought to identify the simplest network capable of solving a musical problem. Comparing the structure of networks trained on different problems provides an indication of their relative complexity. For instance, we discovered that a value unit perceptron could solve the scale tonic problem but could not solve the scale mode problem. This suggests that identifying a scale's tonic is a simpler information-processing problem than identifying its mode. Similarly, the various simulations reported in Chapter 7 indicated that the ii-V-I progression is simpler than the Coltrane changes. This is because an integration device perceptron is all that is required for the former, but a value unit perceptron or a multilayer network of value units is required for the latter, depending upon the choice of encoding.

9.3.4 Network Structure and Composition

The previous sections have pointed out that network interpretations can lead to alternative formal accounts of musical regularities. One interesting possibility raised by this discovery is that the novel formal properties discovered inside a network can be used to provide new methods for musical composition. An example of this possibility is described below.

Atonal music has no discernible musical key or tonal centre because all 12 pitch-classes from Western music occur equally often. Arnold Schoenberg invented a method, called the 12-tone technique or dodecaphony, for composing atonal music. In dodecaphony, one begins a new composition by arranging all 12 pitch-classes in some desired order; this arrangement is the tone row. The first note from the tone row is then used to begin the new piece. The duration of this note, and whether or not it is repeated, is under the composer's control. However, once the use of this note is complete, dodecaphony takes control: the 12-tone method prevents the composer from using it again until all of the other 11 notes in the tone row have been used. Their use, naturally, follows the same procedure used for the first note: the composer decides upon duration and repetition, uses the note, and then moves on to the next note in the tone row. Let us now consider another approach to composing atonal music, one inspired by a feature that we have observed in several network interpretations.

We have seen several examples of artificial neural networks whose hidden units employ connection weights that assign various subsets of pitch-classes to classes, such as the four different circles of major thirds (Figure 6-17) or the two different circles of major seconds (Figure 6-13). Furthermore, these circles are often strange in the sense that the hidden units treat each member of the circle as being the same pitch-class. That is, all of the pitch-classes that belong to one circle of major seconds may be assigned the same connection weight (e.g., to the connection from a pitch-class input unit to a hidden unit). This means that the hidden unit is "deaf" to any differences between members of this subset of pitch-classes. For a hidden unit that uses equivalence classes based on circles of major seconds, there are only two pitch-classes: some "name" x (the weight assigned to C, D, E, F♯, G♯, and A♯) and some other "name" y (the weight assigned to C♯, D♯, F, G, and B).

Why do networks use strange circle equivalence classes to represent musical structure? One reason is that networks discover that notes that belong to the same strange circle are not typically used together to solve musical problems, such as classifying a musical chord. Instead, the network discovers that combining notes from different strange circles is more successful. This use of equivalence classes—combining pitch-classes from different circles, but not from the same circle—suggests an alternative approach to composing atonal music.

Imagine a musical composition constructed from a set of different musical voices. Let each of these voices be derived from one strange circle. The notes sung by this voice are selected by randomly choosing from the set of pitch-classes that belong to the strange circle. For instance, if one voice was associated with a particular circle of major thirds, then one could write its notes by randomly choosing one note at a time from the set [C, E, G♯]. To make the voice more musically interesting, one could add a randomly selected rest to the mix by selecting from the set [C, E, G♯, R] where R indicates a rest (i.e., no note is to be sung).

If one associated different voices with different strange circles, and composed via random selection as described above, then one would be following the general principle discovered by the network: pitch-classes from different strange circles can occur together, but pitch-classes from the same strange circle cannot. Furthermore, one could use this method to compose atonal music by wisely choosing which strange circles to use to create different voices. For instance, imagine creating a piece that included four voices, each associated with a different circle of major thirds. This composition would be atonal, in Schoenberg's sense, because the four circles combine to include all 12 possible pitch-classes. Randomly selecting pitches from each of these circles would produce a composition that did not have a tonal centre because each of the 12 pitch-classes would occur equally often when the composition was considered as a whole.

One example of this sort of composition can be found at the following website: http://cognitionandreality.blogspot.ca/2013/03/composing-atonal-music-using-strange.html. The website provides a musical score created by using this approach to composition. This score includes six staves, one for each voice. Each voice is generated by randomly selecting from one strange circle (and including rests in this sampling procedure). The top two staves, written in quarter notes, are each drawn from a different circle of major seconds. The bottom four staves, written in half notes, are each drawn from a different circle of major thirds. The score is created by applying two additional musical assumptions. First, while each wheel generates a pitch-class name, the composer decided how high or low (in terms of octave) each note is positioned. Second, in order to ensure that all notes tend to occur equally often in the score, the two circles of major seconds are sampled twice as frequently relative to the other four strange circles.

At the bottom of this web page, one can find links that play some of the voices individually, some combinations of a small number of the voices, and all of the voices together. On listening to these samples, one discovers that individual voices or strange circles are musical, but are not musically interesting. Music that is more interesting emerges from combining the random outputs of different circles. Other sets of strange circles than those used to create the score discussed above could also be used for composing. What kinds of atonal pieces can be created when many different strange circles are available? To answer this question, I created a Java program that uses David Koelle's music package jFugue (Koelle, 2008). This package lets the programmer define strings of musical notes, and then takes care of playing them. The program that I wrote lets the user choose a composition's tempo and length with a mouse, and then make a checkmark beside every strange circle to be used in a piece. The user can decide whether to include rests, and set the duration and the octave (2 is lowest, 5 is highest) for each set of circles. A press of the "compose" button leads to a pause while the various voices are constructed, and then the piece is played through the computer's speakers. One can easily explore the possibilities of strange circle composing with this program and listening to the sounds that it creates. This program is also available as part of the same blog post mentioned above.

9.4 Implications for Musical Cognition

Music theory, and its formalizations, defines the input/output mappings that our artificial neural networks have learned. Thus, it provides the vocabulary for the computational level analysis of the networks. My network interpretations have revealed how these computational mappings are mediated, and thus provide the algorithmic level analysis. However, we saw in Section 9.3 that these algorithmic level results could inform the computational level as well. What are the implications of our simulations for the study of musical cognition?

In general, the experimental study of cognition focuses upon the algorithmic level. This is because experimental cognitive psychology attempts to discover the procedures used by human subjects to process information (Dawson, 1998, 2013). From this perspective, there should be an important relationship between results in musical cognition and our simulations.

9.4.1 Networks and Algorithms

Even the most committed forward engineer realizes that at some point their models must be related to human data. A synthetic cognitive science of music must eventually find empirical links between networks and human musical cognition. How are these links to be established? Fortunately, when networks are trained they provide a great deal of different kinds of evidence that can be used to compare their musical representations and processes with those of human subjects.

To illustrate, let us consider what is called relative complexity evidence (Pylyshyn, 1984). Relative complexity evidence compares a system's processing of one type of stimulus to another. For instance, when training musical networks, this could involve comparing the learning of different patterns over time. Are some types of patterns harder to learn than other patterns?

In Chapter 6, I interpreted the structure of a multilayer perceptron trained to classify four different types of triads. To collect relative complexity evidence, I could save the state of the network (e.g., its structure, its responses to patterns, its errors) after every 250 epochs of training.

When this is done, some interesting properties are revealed. A typical network of this type performs well on augmented and diminished triads very early in training. It generates highly accurate responses to both types of these triads after only 250 epochs of training. In contrast, it has more difficulty learning both major and minor triads. About 1000 epochs of training are required to reduce the error generated to major triads to the same level of error generated to the augmented and diminished triads. Minor triads provide the greatest challenge; about 1750 sweeps of training are required before this type of stimulus is learned.

Relative complexity evidence can be easily obtained to get a sense of the dynamics of learning a particular musical task. How might we compare this evidence to the behaviour of human subjects? One approach is to create a new experimental paradigm, one that is as straightforward as the probe tone method (Krumhansl, 1990a). In general, this new experimental paradigm involves teaching human subjects on the same musical task that was presented to the network, where this teaching is done in a manner similar to that used for network training. For example, consider the triad classification problem from Chapter 6. One can build a block of training patterns to present a human subject, where a single block includes each of the 48 triads. In a given block, the order of patterns is randomized. During training, a subject hears a triad, and then classifies the sound. For instance, they might assign the sound to one of four different categories (A, B, C, or D). Of course, each category is associated with a triad type, but the subject need not be provided with the names of these types in order to respond. Because the subject is being trained in a fashion analogous to the network, learning needs to be supervised. After the subject classifies the stimulus, they are told what the correct response was. Then the next stimulus in the block can be presented.

Note that in this paradigm a block of trials for a human subject is analogous to an epoch of training for a network. So, if a subject is run through a series of training blocks, then they are learning in a similar fashion to the network. We continue training until an acceptable degree of accuracy has been achieved, assuming that the feedback that a subject receives after each trial improves their performance. Once the subject has "converged" to a solution to the triad classification problem, their data can be analyzed in a fashion similar to that of the network's. For instance, the subject's average accuracy to each triad type can be measured for each block of training. As a result, relative complexity evidence for human subjects can be directly compared to the same kind of evidence collected for networks. One could argue that the similarity between these two sources of data reflects the degree that the network is learning the problem in the same way as a human.

Of course, in practice this paradigm would involve exploring a number of different design decisions. We saw earlier that one could use different encodings of the same training set to affect network learning. Clearly, a variety of network encodings would have to be explored and compared to human data. We also have a variety of design decisions to explore when the human data is collected. Is performance affected by the timbre of stimuli? Is it affected by the octave in which stimuli are presented? Is it affected by inversions of chords? In short, this general approach to studying musical cognition opens the door to a wide range of studies that involve exploring different tasks, different network settings, and different experimental stimuli.

9.4.2 Musical Representations

The discussion in Section 9.4.1 concerns how one might use artificial neural networks to inform the algorithmic analysis of musical cognition, and how to explore the relationship between how networks and humans learn the same musical task. However, in addition to specifying algorithms cognitive scientists also must specify the architecture of cognition (Dawson, 1998, 2013). That is, they must determine the basic representations and operations available for solving an information-processing problem (Pylyshyn, 1984). Interpreting musical networks can inform the architectural mission of the cognitive science of music.

As was noted in Chapter 1, one of the central assumptions of cognitivism is that humans actively process information. Musical cognition is thought to proceed by actively integrating musical stimuli with mental representations of music (Cook, 1999; Deliège & Sloboda, 1997; Deutsch, 1982, 1999, 2013; Francès, 1988; Howell et al., 1985; Krumhansl, 1990a; Lerdahl, 2001; Lerdahl & Jackendoff, 1983; Sloboda, 1985; Snyder, 2000; Temperley, 2001). Furthermore, the act of organizing the music that we hear can affect how we represent it; presumably, musical representations change as a function of our musical experience and training.

It is therefore not surprising that at the heart of musical cognition one finds proposals about the nature of musical representation. For instance, the evidence supporting the existence of the tonal hierarchy (Krumhansl, 1990a) suggests that musical harmony is represented hierarchically in a system that makes certain musical structures more stable, central, or important than others depending upon context (Bharucha & Krumhansl, 1983). This in turn suggests that musical representation may be analogous to how semantic concepts are represented in prototype theory (Rosch, 1975; Rosch & Mervis, 1975).

Other kinds of representation have been proposed. The tonal hierarchy for each key could be explicitly represented, as is required in models of key-finding (Krumhansl & Kessler, 1982). Harmonic structures could be represented spatially, where distances between represented entities reflect their similarity (Krumhansl, 2005). There is a long tradition of employing spatial manifolds as representational primitives for cognition (Cutting, 1986; Cutting & Proffitt, 1982; Kosslyn, 1980, 1994; Shepard, 1984, 1990). Some have proposed that musical cognition is mediated by a language-like generative grammar (Lerdahl, 2001; Lerdahl & Jackendoff, 1983), while others have proposed representations that capture music's probabilistic structure (Huron, 2006; Temperley, 2007). In the context of such representational proposals, what is the role of the simulations that we explored in preceding chapters?

If a key goal of musical cognitivism is to identify potential representational formats, then this search should be as broad as possible. Artificial neural networks offer a medium for discovering new representational proposals. Many of our discoveries that were summarized earlier in this chapter—the prominent tritone, strange circles, and coarse codes—can be interpreted not just as contributions to music theory but also as contributions to music cognition. When we pull these regularities out of our networks, it is reasonable to ask whether they might also play a role in human cognition.

A network interpretation might simply point to musical information that a representation should make explicit because it affects musical information processing. For instance, we saw that the tritone plays an important role in many of our networks. This is consistent with some results in the psychology of music. It has been claimed that human listeners can quickly identify a composition's musical key by detecting the presence of rare musical intervals like the tritone (Brown & Butler, 1981; Browne, 1981; Butler, 1989; Van Egmond & Butler, 1997), although this theory has not gone unchallenged (Krumhansl, 1990b). Perhaps, more importantly, our network interpretations may also inform architectural proposals for musical cognition. When I discover a particular representational structure in my networks, like the use of particular strange circles, or a specific kind of coarse coding, it is natural to ask whether these structures are also part of the cognitive architecture for music. For instance, are strange circles literally part of the structure of musical representations? To answer such a question, one must design experiments that explore the relevance of the proposed representation. However, designing these studies comes after generating the question. An important source of such questions is network interpretation.

9.5 Future Directions

This book explored the viability of a synthetic cognitive science of music. Simple artificial neural networks learned basic musical tasks related to Western tonality. When a network completed its training, I interpreted its internal structure. The primary question explored in this book is whether this approach can inform the cognitive science of music. The results summarized in the current chapter, and detailed in the preceding chapters, should convince the reader that the approach introduced in this book holds a great deal of promise. Even though I adopted a very simple approach to training our networks, and even though I trained networks on tasks that are well understood in traditional music theory, I was able to uncover a number of novel and surprising results. My network interpretations revealed a number of representational insights that can inform both music theory (Section 9.3) and musical cognition (Section 9.4). Now that I have established the viability of this approach, it is possible use it to venture into further, more complex, domains.

With respect to computational-level investigations, researchers can now proceed to explore a greater variety of musical information-processing problems. For instance, the Forte numbering system from musical set theory is used to assign an identifying number to a musical entity (Forte, 1973, 1985). This is useful because in many cases two musical entities that seem to be quite different may be assigned the same Forte number. This in turn implies that they are functionally equivalent. This means that musical set theory can define problems that are more involved where, for example, two musical entities that are assigned the same Forte number generate identical network outputs.

With respect to algorithmic-level investigations, experimental psychologists can now proceed to research that has the goal of relating evidence gathered from networks to analogous evidence collected from human subjects. A general approach to this kind of research was proposed in Section 9.4. Other wider variations are also feasible. For instance, one could train neural networks on a task analogous to Krumhansl's probe tone method. One could also train networks to rate the dissonance or consonance of musical stimuli where network output is informed not by music theory but instead by existing experimental results.

With respect to the architectural level, or even the implementational level, researchers can now consider more complex sorts of encodings. All of the networks reported in this book have encoded stimuli in a fashion that maps directly onto music theory (e.g., pitch-class, pitch). Other physical or physiological encodings are possible. For example, what is the effect of representing musical inputs as collections of sine wave frequencies, or in a fashion that emulates the encoding of the basilar membrane?

Similarly, the design decisions that guided architectural selections for my simulations led me to focus on musical regularities that could be obtained from spatially represented stimuli. For the most part, I avoided the study of musical stimuli that were presented through time, or the study of temporal musical regularities such as rhythm. My success in the simulations reported in this book indicates that studying temporal aspects of music with (interpreted) networks is a crucial next step. Furthermore, the exploration of music is not limited to the architectures that are employed in this book; many other network architectures can be explored (Griffith & Todd, 1999; Todd & Loy, 1991). These include self-organizing networks that learn statistical properties of inputs without requiring an external teacher (Gjerdingen, 1990; Kohonen, 2001; Page, 1994), or recurrent networks that are explicitly designed to detect regularities in time (Elman, 1990; Franklin, 2004, 2006).

There is also a growing interest in deep learning networks (Bengio et al., 2013; Hinton, 2007; Hinton et al., 2006; Hinton & Salakhutdinov, 2006; Larochelle et al., 2012). These networks, which have many more layers of hidden units than we have been considering in this book, have solved many difficult pattern recognition tasks in natural language, image classification, and the processing of sound (Hinton, 2007; Hinton et al., 2006; Mohamed, Dahl, & Hinton, 2012; Sarikaya, Hinton, & Deoras, 2014). Some of these tasks involve processing music, including its temporal properties (Humphrey, Bello, & LeCun, 2013).

The power of deep learning as a technology is becoming well established. However, it is important to remember that the goal of a connectionist cognitive science of music is not to generate new technologies. Instead, it is to enhance our understanding of musical cognition, or of music theory, by providing insights into these domains. These insights require us to investigate how networks solve problems, and to use these interpretations of network processing to inform theory. Deep learning provides a powerful technology, but techniques for interpreting the structure of deep belief networks are in their infancy (Erhan et al., 2010). Until their internal structure can be fundamentally understood, these powerful networks are likely not going to provide new directions to a cognitive science of music.

My hope is that the results reported in this book will serve as an impetus for continued exploration, pursuing investigations of additional musical properties using the architectures described here, or employing new kinds of artificial neural networks. However, it is important to remember that the success of a connectionist cognitive science of music depends on one fundamental research goal: interpreting the internal structure of a network after it learns. Network interpretations will be the source of new theoretical insights into musical cognition.

References

Ackley, D. H., Hinton, G. E., & Sejnowski, T. J. (1985). A learning algorithm for Boltzman machines. *Cognitive Science*, 9, 147–169.

Adiloglu, K., & Alpaslan, F. N. (2007). A machine learning approach to two-voice counterpoint composition. *Knowledge-Based Systems*, *20*(3), 300–309. doi: 10.1016/j.knosys.2006.04.018

Albrecht, J. D., & Shanahan, D. (2013). The use of large corpora to train a new type of key-finding algorithm: An improved treatment of the minor mode. *Music Perception*, *31*(1), 59–67. doi: 10.1525/mp.2013.31.1.59

Allen, D. (1967). Octave discriminability of musical and non-musical subjects. *Psychonomic Science*, *7*, 421–422.

Amit, D. J. (1989). *Modeling brain function: The world of attractor neural networks*. Cambridge, UK: Cambridge University Press.

Anderson, J. A. (1995). *An introduction to neural networks*. Cambridge, MA: MIT Press.

Atkinson, R. C., Bower, G. H., & Crothers, E. J. (1965). *An introduction to mathematical learning theory*. New York, NY: John Wiley & Sons.

Babbitt, M. (1960). Twelve-tone invariants as compositional determinants. *The Musical Quarterly*, *46*(2), 246–259.

Babbitt, M. (1961). Set structure as a compositional determinant. *Journal of Music Theory*, *5*(1), 72–94.

Baesens, B., Setiono, R., Mues, C., & Vanthienen, J. (2003). Using neural network rule extraction and decision tables for credit-risk evaluation. *Management Science*, *49*(3), 312–329. doi: 10.1287/mnsc.49.3.312.12739

Ballard, D. (1986). Cortical structures and parallel processing: Structure and function. *Journal of Music Theory*, *21*(2), 293–323.

Barbour, J. M. (1972). *Tuning and temperament: A historical survey*. New York, NY: Da Capo Press.

Barlow, H. B. (1972). Single units and sensation: A neuron doctrine for perceptual psychology? *Perception*, *1*, 371–394.

Barlow, H.B. (1995). The neuron doctrine in perception. In M. S. Gazzaniga (Ed.), *The cognitive neurosciences* (pp. 415–435). Cambridge, MA: MIT Press.

Bechtel, W. (1994). Natural deduction in connectionist systems. *Synthese*, *101*, 433–463.

Bechtel, W., & Abrahamsen, A. A. (2002). *Connectionism and the mind: Parallel processing, dynamics, and evolution in networks* (2nd ed.). Malden, MA: Blackwell.

Bellgard, M. I., & Tsang, C. P. (1994). Harmonizing music the Boltzmann way. *Connection Science, 6*, 281–297.
Bengio, Y., Courville, A., & Vincent, P. (2013). Representation learning: A review and new perspectives. *IEEE Transactions on Pattern Analysis and Machine Intelligence, 35*(8), 1798–1828. doi: 10.1109/tpami.2013.50
Benuskova, L. (1994). Modeling the effect of the missing fundamental with an attractor neural network. *Network: Computation in Neural Systems, 5*(3), 333–349.
Benuskova, L. (1995). Modeling transpositional invariancy of melody recognition with an attractor neural network. *Network: Computation in Neural Systems, 6*(3), 313–331.
Berkeley, I., & Raine, R. (2011). An old-fashioned connectionist approach to a Cajun chord change problem. *Connection Science, 23*(3), 209–218. doi: 10.1080/09540091.2011.597500
Berkeley, I. S. N., Dawson, M. R. W., Medler, D. A., Schopflocher, D. P., & Hornsby, L. (1995). Density plots of hidden value unit activations reveal interpretable bands. *Connection Science, 7*, 167–186.
Berkeley, I. S. N., & Gunay, C. (2004). Conducting banding analysis with trained networks of sigmoid units. *Connection Science, 16*(2), 119–128. doi: 10.1080/09540090412331282278
Berkowitz, A.L. (2010). *The improvising mind: Cognition and creativity in the musical moment*. New York, NY: Oxford University Press.
Bertalanffy, L.v. (1967). *Robots, men, and minds*. New York, NY: G. Braziller.
Bertalanffy, L.v. (1969). *General system theory: Foundations, development, applications*. New York, NY: G. Braziller.
Bharucha, J., & Krumhansl, C. L. (1983). The representation of harmonic structure in music: Hierarchies of stability as a function of context. *Cognition, 13*(1), 63–102. doi: 10.1016/0010-0277(83)90003tw
Bharucha, J. J. (1987). Music cognition and perceptual facilitation: A connectionist framework. *Music Perception, 5*(1), 1–30.
Bharucha, J. J. (1999). Neural nets, temporal composites, and tonality. In D. Deutsch (Ed.), *The psychology of music* (2nd ed., pp. 413–440). San Diego, CA: Academic Press.
Bharucha, J. J., & Todd, P. M. (1989). Modeling the perception of tonal structure with neural nets. *Computer Music Journal, 13*(4), 44–53.
Bidelman, G. M., & Krishnan, A. (2009). Neural correlates of consonance, dissonance, and the hierarchy of musical pitch in the human brainstem. *Journal of Neuroscience, 29*(42), 13165–13171. doi: 10.1523/jneurosci.3900-09.2009
Bishop, C. M. (1995). *Neural networks for pattern recognition*. New York, NY: Oxford University Press.
Blackwell, H. R., & Schlosberg, H. (1943). Octave generalization, pitch discrimination, and loudness thresholds in the white rat. *Journal of Experimental Psychology, 33*(5), 407–419. doi: 10.1037/h0057863

Boole, G. (1854/2003). *The laws of thought*. Amherst, NY: Prometheus Books (Originally published in 1854).

Boring, E. G. (1950). *A history of experimental psychology*. New York, NY: Appleton-Century-Crofts.

Braitenberg, V. (1984). *Vehicles: Explorations in synthetic psychology*. Cambridge, MA: MIT Press.

Brown, H., & Butler, D. (1981). Diatonic trichords as minimal tonal cue-cells. *In Theory Only*, *5*(6 & 7), 37–55.

Browne, R. (1981). Tonal implications of the diatonic set. *In Theory Only*, *5*(6–7), 3–21.

Broze, Y., & Shanahan, D. (2013). Diachronic changes in jazz harmony: A cognitive perspective. *Music Perception*, *31*(1), 32–45. doi: 10.1525/mp.2013.31.1.32

Bruhn, S. (2014). *J. S. Bach's well-tempered clavier: In-depth analysis and interpretation*. Waldkirch, Germany: Edition Gorz.

Bruner, J. S. (1990). *Acts of meaning*. Cambridge, MA: Harvard University Press.

Bugatti, A., Flammini, A., & Migliorati, P. (2002). Audio classification in speech and music: A comparison between a statistical and a neural approach. *Eurasip Journal on Applied Signal Processing*, *2002*(4), 372–378.

Burnod, Y. (1990). *An adaptive neural network: The cerebral cortex*. London UK: Prentice-Hall.

Buus, S., Lauemoller, S.L., Worning, P., Kesmir, C., Frimurer, T., Corbet, S., . . . Brunak, S. (2003). Sensitive quantitative predictions of peptide-MHC binding by a 'Query by Committee' artificial neural network approach. *Tissue Antigens*, *62*(5), 378–384. doi: 10.1034/j.1399-0039.2003.00112.x

Butler, D. (1989). Describing the perception of tonality in music: A critique of the tonal hierarchy theory and a proposal for a theory of intervallic rivalry. *Music Perception*, *6*(3), 219–242.

Cangelosi, A. (2010). Connectionist modelling of music emotions. *Physics of Life Reviews*, *7*(1), 37–38. doi: 10.1016/j.plrev.2010.01.005

Carpenter, G. A., & Grossberg, S. (1992). *Neural networks for vision and image processing*. Cambridge, MA: MIT Press.

Caudill, M., & Butler, B. (1992). *Understanding neural networks* (Vol.1). Cambridge MA: MIT Press.

Chalmers, J. (1992). *Divisions of the tetrachord: A prolegomenon to the construction of musical scales*. Lebanon, NH: Frog Peak Music.

Chomsky, N. (1965). *Aspects of the theory of syntax*. Cambridge, MA: MIT Press.

Churchland, P. S. (1986). *Neurophilosophy*. Cambridge, MA: MIT Press.

Churchland, P. S., & Sejnowski, T. J. (1992). *The computational brain*. Cambridge, MA: MIT Press.

Clark, A. (1993). *Associative engines*. Cambridge, MA: MIT Press.

Claudon, F. (1980). *The concise encyclopedia of Romanticism*. Secaucus, NJ: Chartwell Books.

Cohen, H. F. (1984). *Quantifying music: The science of music at the first stage of the Scientific Revolution, 1580–1650*. Boston, MA: D. Reidel.

Conrad, R. (1964). Information, acoustic confusion, and memory span. *British Journal of Psychology*, *55*, 429–432.

Cook, P. R. (1999). *Music, cognition and computerized sound*. Cambridge, MA: MIT Press.

Coombs, C. H., Dawes, R. M., & Tversky, A. (1970). *Mathematical psychology: An elementary introduction*. Englewood Cliffs, NJ: Prentice-Hall.

Coutinho, E., & Cangelosi, A. (2009). The use of spatio-temporal connectionist models in psychological studies of musical emotions. *Music Perception*, *27*(1), 1–15. doi: 10.1525/mp.2009.27.1.1

Creighton, H. (1932). *Songs and ballads from Nova Scotia*. Toronto, ON: J. M. Dent.

Cummins, R. (1983). *The nature of psychological explanation*. Cambridge, MA: MIT Press.

Cutting, J. E. (1986). *Perception with an eye for motion*. Cambridge, MA: MIT Press.

Cutting, J. E., & Proffitt, D. (1982). The minimum principle and the perception of absolute, common, and relative motions. *Cognitive Psychology*, *14*, 211–246.

Cynx, J. (1993). Auditory frequency generalization and a failure to find octave generalization in a songbird, the European starling (sturnus vulgaris). *Journal of Comparative Psychology*, *107*(2), 140–146. doi: 10.1037/0735-7036.107.2.140

Das, A., Reddy, N. P., & Narayanan, J. (2001). Hybrid fuzzy logic committee neural networks for recognition of swallow acceleration signals. *Computer Methods and Programs in Biomedicine*, *64*(2), 87–99. doi: 10.1016/s0169-2607(00)00099-7

Dawson, M. R. W. (1998). *Understanding cognitive science*. Oxford, UK: Blackwell.

Dawson, M. R. W. (2004). *Minds and machines: Connectionism and psychological modeling*. Malden, MA: Blackwell.

Dawson, M. R. W. (2005). *Connectionism: A Hands-on Approach*. Oxford, UK: Blackwell.

Dawson, M. R. W. (2008). Connectionism and classical conditioning. *Comparative Cognition and Behavior Reviews*, *3* (Monograph), 1–115.

Dawson, M. R. W. (2009). Computation, cognition—and connectionism. In D. Dedrick & L. Trick (Eds.), *Cognition, computation, and Pylyshyn* (pp. 175–199). Cambridge, MA: MIT Press.

Dawson, M. R. W. (2013). *Mind, body, world: Foundations of cognitive science*. Edmonton, AB: Athabasca University Press.

Dawson, M. R. W., & Boechler, P. M. (2007). Representing an intrinsically nonmetric space of compass directions in an artificial neural network. *International Journal of Cognitive Informatics and Natural Intelligence*, *1*, 53–65.

Dawson, M. R. W., Boechler, P. M., & Orsten, J. (2005). An artificial neural network that uses coarse allocentric coding of direction to represent distances between locations in a metric space. *Spatial Cognition and Computation*, 5, 29–57.

Dawson, M. R.W., Boechler, P. M., & Valsangkar-Smyth, M. (2000a). Representing space in a PDP network: Coarse allocentric coding can mediate metric and nonmetric spatial judgements. *Spatial Cognition and Computation*, 2, 181–218.

Dawson, M. R.W., & Dupuis, B. (2012). Equilibria of perceptrons for simple contingency problems. *IEEE Transactions on Neural Networks and Learning Systems*.

Dawson, M. R. W., Dupuis, B., Spetch, M. L., & Kelly, D. M. (2009). Simple artificial networks that match probability and exploit and explore when confronting a multiarmed bandit. *IEEE Transactions on Neural Networks*, *20*(8), 1368–1371.

Dawson, M. R. W., Dupuis, B., & Wilson, M. (2010a). *From bricks to brains: The embodied cognitive science of LEGO robots*. Edmonton, AB: Athabasca University Press.

Dawson, M. R. W., & Gupta, M. (2017). Probability matching in perceptrons: Effects of conditional dependence and linear nonseparability. *PloS ONE*, *12*(2), e0172431. doi: doi:10.1371/journal.pone.0172431

Dawson, M. R. W., Kelly, D. M., Spetch, M. L., & Dupuis, B. (2010b). Using perceptrons to explore the reorientation task. *Cognition*, *114*(2), 207–226.

Dawson, M. R. W., Medler, D. A., & Berkeley, I. S. N. (1997). PDP networks can provide models that are not mere implementations of classical theories. *Philosophical Psychology*, *10*, 25–40.

Dawson, M. R. W., Medler, D. A., McCaughan, D. B., Willson, L., & Carbonaro, M. (2000b). Using extra output learning to insert a symbolic theory into a connectionist network. *Minds and Machines*, *10*, 171–201.

Dawson, M. R. W., & Piercey, C. D. (2001). On the subsymbolic nature of a PDP architecture that uses a nonmonotonic activation function. *Minds and Machines*, *11*, 197–218.

Dawson, M. R. W., & Schopflocher, D. P. (1992). Modifying the generalized delta rule to train networks of nonmonotonic processors for pattern classification. *Connection Science*, *4*, 19–31.

Dawson, M. R. W., & Shamanski, K. S. (1994). Connectionism, confusion and cognitive science. *Journal of Intelligent Systems*, *4*, 215–262.

Dawson, M. R. W., & Zimmerman, C. (2003). Interpreting the internal structure of a connectionist model of the balance scale task. *Brain & Mind*, *4*, 129–149.

Deliège, I., & Sloboda, J.A. (1997). *Perception and cognition of music*. Hove, East Sussex, UK: Psychology Press.

Demany, L., & Armand, F. (1984). The perceptual reality of tone chroma in early infancy. *Journal of the Acoustical Society of America*, *76*(1), 57–66. doi: 10.1121/1.391006

Demsey, D. (1991). Chromatic third relations in the music of John Coltrane. *Annual Review Of Jazz Studies*, 5, 145–180.
Desain, P., & Honing, H. (1989). The quantization of musical time: A connectionist approach. *Computer Music Journal, 13*(3), 56–66.
Descartes, R. (1637/2006). *A discourse on the method of correctly conducting one's reason and seeking truth in the sciences* (I. Maclean, Trans.). New York, NY: Oxford University Press.
Descartes, R. (1641/1996). *Meditations on first philosophy* (Rev. ed.). New York, NY: Cambridge University Press.
Deutsch, D. (1982). *The psychology of music*. New York, NY: Academic Press.
Deutsch, D. (1986). A musical paradox. *Music Perception, 3*(3), 275–280.
Deutsch, D. (1987). The tritone paradox: Effects of spectral variables. *Perception & Psychophysics, 41*(6), 563–575. doi: 10.3758/bf03210490
Deutsch, D. (1991). The tritone paradox: An influence of language on music perception. *Music Perception, 8*(4), 335–347.
Deutsch, D. (1999). *The psychology of music* (2nd ed.). San Diego, CA: Academic Press.
Deutsch, D. (2010). The paradox of pitch circularity. *Acoustics Today, 6*(3), 8–15.
Deutsch, D. (2013). *The psychology of music* (3rd ed.). Waltham, MA: Academic Press.
Deutsch, D., & Boulanger, R.C. (1984). Octave equivalence and the immediate recall of pitch sequences. *Music Perception, 2*(1), 40–51.
DeVito, C., & Porter, L. (2008). *The John Coltrane reference*. New York, NY: Routledge.
Dhombres, J. (2002). Lagrange, "working mathematician," on music considered as a source for science. In G. Assayag, H. G. Feichtinger & J.-F. Rodrigues (Eds.), *Mathematics and Music* (pp. 65–78). New York, NY: Springer.
Donahue, T. (2005). *A guide to musical temperament*. Lanham, MD.: Scarecrow Press.
Dourish, P. (2001). *Where the action is: The foundations of embodied interaction*. Cambridge, MA: MIT Press.
Dreyfus, H. L. (1972). *What computers can't do: A critique of artificial reason* (1st ed.). New York, NY: Harper & Row.
Dreyfus, H. L. (1992). *What computers still can't do*. Cambridge, MA: MIT Press.
Duch, W., & Jankowski, N. (1999). Survey of neural transfer functions. Neural *Computing Surveys*, 2, 163–212.
Dutton, J. M., & Starbuck, W. H. (1971). *Computer simulation of human behavior*. New York, NY: John Wiley & Sons.
Ede, A., & Cormack, L. B. (2004). *A history of science in society: From philosophy to utility*. Peterborough, ON: Broadview Press.
Einstein, A. (1947). *Music in the Romantic era*. New York: W. W. Norton.
Elman, J. (1990). Finding structure in time. *Cognitive Science, 14*, 179–211.
Enquist, M., & Ghirlanda, S. (2005). *Neural networks and animal behavior*. Princeton, NJ: Princeton University Press.

Erhan, D., Courville, A., & Bengio, Y. (2010). *Understanding representations learned in deep architectures* (Technical Report 1355). Departement d'Informatique et Recherche Operationnelle, Université de Montréal.

Farrell, J. E., & Shepard, R. N. (1981). Shape, orientation, and apparent rotational motion. *Journal of Experimental Psychology: Human Perception and Performance*, *7*, 477–486.

Fechner, G. T. (1966/1860). *Elements of psychophysics* (H. E. Adler, Trans., D. H. Howes & E. G. Boring Eds.). New York, NY: Holt.

Feigenbaum, E. A., & Feldman, J. (1995). *Computers and thought*. Cambridge, MA: MIT Press.

Fiske, H. E. (2004). *Connectionist models of musical thinking*. Lewiston, NY: E. Mellen Press.

Fletcher, H. (1924). The physical criterion for determining the pitch of a musical tone. *Physical Review*, *23*(3), 427–437.

Fodor, J. A. (1975). *The language of thought*. Cambridge, MA: Harvard University Press.

Forte, A. (1973). *The structure of atonal music*. New Haven, CT: Yale University Press.

Forte, A. (1985). Pitch-class set analysis today. *Music Analysis*, *4*(1–2), 29–58. doi: 10.2307/854234

Francès, R. (1988). *The perception of music*. Hillsdale, N.J.: L. Erlbaum.

Frankland, B. W., & Cohen, A. J. (1996). Using the Krumhansl and Schmuckler key-finding algorithm to quantify the effects of tonality in the interpolated-tone pitch-comparison task. *Music Perception*, *14*(1), 57–83.

Franklin, J. A. (2004). Recurrent neural networks and pitch representations for music tasks. Paper presented at the Seventeenth International Florida Artificial Intelligence Research Symposium Conference, Miami Beach, Florida.

Franklin, J. A. (2006). Jazz melody generation using recurrent networks and reinforcement learning. *International Journal on Artificial Intelligence Tools*, *15*(4), 623–650.

Gaines, J. R. (2005). *Evening in the palace of reason: Bach meets Frederick the Great in the Age of Enlightenment*. New York, NY: Fourth Estate.

Gallant, S. I. (1993). *Neural network learning and expert systems*. Cambridge, MA: MIT Press.

Gasser, M., Eck, D., & Port, R. (1999). Meter as mechanism: A neural network model that learns metrical patterns. *Connection Science*, *11*(2), 187–216.

Gjerdingen, R. O. (1990). Categorization of musical patterns by self-organizing neuron-like networks. *Music Perception*, *7*(4), 339–369.

Gjerdingen, R. O. (1992). Learning syntactically significant temporal patterns of chords: A masking field embedded in an ART-3 architecture. *Neural Networks*, *5*(4), 551–564.

Gjerdingen, R. O., & Perrott, D. (2008). Scanning the dial: The rapid recognition of music genres. *Journal of New Music Research, 37*(2), 93–100. doi: 10.1080/09298210802479268

Gluck, M. A., & Myers, C. (2001). *Gateway to memory: An introduction to neural network modeling of the hippocampus and learning*. Cambridge, MA: MIT Press.

Graham, R., & Dawson, M. (2005). Using artificial neural networks to examine event-related potentials of face memory. *Neural Network World, 15*, 215–227.

Griffith, N. (1995). Connectionist visualization of tonal structure. *Artificial Intelligence Review, 8*(5–6), 393–408.

Griffith, N., & Todd, P. M. (1999). *Musical networks: Parallel distributed perception and performance*. Cambridge, MA: MIT Press.

Grossberg, S. (1980). How does the brain build a cognitive code? *Psychological Review, 87*, 1–51.

Grossberg, S. (1987). Competitive learning: From interactive activation to adaptive resonance. *Cognitive Science, 11*, 23–63.

Grossberg, S. (1988). *Neural networks and natural intelligence*. Cambridge, MA: MIT Press.

Gu, H., & Lin, Z. (2014). Singing-voice synthesis using ANN vibrator-parameter models. *Journal of Information Science and Engineering, 30*(2), 425–442.

Guernsey, M. (1928). The role of consonance and dissonance in music. *American Journal of Psychology, 40*, 173–204. doi: 10.2307/1414484

Guo, J. J., & Luh, P. B. (2004). Improving market clearing price prediction by using a committee machine of neural networks. *IEEE Transactions on Power Systems, 19*(4), 1867–1876. doi: 10.1109/tpwrs.2004.837759

Handelman, E. J., & Sigler, A. (2013). Key induction and key mapping using pitch-class set assertions. In J. Yust, J. Wild, & J. A. Burgoyne (Eds.), *Mathematics and computation in music* (pp. 115–127). New York, NY: Springer.

Hanslick, E. (1854/1957). *The beautiful in music*. New York, NY: Liberal Arts Press.

Hanson, H. (1960). *Harmonic materials of modern music*. New York, NY: Appleton-Century-Crofts.

Hanson, S. J., & Burr, D. J. (1990). What connectionist models learn: Learning and representation in connectionist networks. *Behavioral and Brain Sciences, 13*, 471–518.

Hayashi, Y., Setiono, R., & Yoshida, K. (2000). A comparison between two neural network rule extraction techniques for the diagnosis of hepatobiliary disorders. *Artificial Intelligence in Medicine, 20*(3), 205–216. doi: 10.1016/s0933-3657(00)00064-6

Heidegger, M. (1927/1962). *Being and time* (J. Macquarrie and E. Robinson, Trans.). New York, NY: Harper & Row.

Helmholtz, H., & Ellis, A. J. (1863/1954). *On the sensations of tone as a physiological basis for the theory of music* (2nd English ed.). New York, NY: Dover Publications.

Hiebert, E. (2014). *The Helmholtz legacy in physiological acoustics*. London, UK: Springer.

Hinton, G. E. (1986). Learning distributed representations of concepts. Paper presented at the 8th Annual Meeting of the Cognitive Science Society, Ann Arbor, MI.

Hinton, G. E. (2007). Learning multiple layers of representation. *Trends in Cognitive Sciences, 11*(10), 428–434. doi: 10.1016/j.tics.2007.09.004

Hinton, G. E., McClelland, J., & Rumelhart, D. (1986). Distributed representations. In D. Rumelhart & J. McClelland (Eds.), *Parallel distributed processing* (Vol. 1, pp. 77–109). Cambridge, MA: MIT Press.

Hinton, G. E., Osindero, S., & Teh, Y. (2006). A fast learning algorithm for deep belief nets. *Neural Computation, 18*(7), 1527–1554. doi: 10.1162/neco.2006.18.7.1527

Hinton, G. E., & Salakhutdinov, R. R. (2006). Reducing the dimensionality of data with neural networks. *Science, 313*(5786), 504–507. doi: 10.1126/science.1127647

Hoeschele, M., Weisman, R. G., Guillette, L. M., Hahn, A. H., & Sturdy, C. B. (2013). Chickadees fail standardized operant tests for octave equivalence. *Animal Cognition, 16*(4), 599–609. doi: 10.1007/s10071-013-0597-z

Hofstadter, D. R. (1979). *Godel, Escher, Bach: An eternal golden braid*. New York, NY: Basic Books.

Holtzman, S. R. (1977). Program for key determination. *Interface: Journal of New Music Research, 6*(1), 29–56.

Hook, J. L. (2006). Exploring musical space. *Science, 313*(5783), 49–50. doi: 10.1126/science.1129300

Hoover, A. K., & Stanley, K. O. (2009). Exploiting functional relationships in musical composition. *Connection Science, 21*(2–3), 227–251. doi: 10.1080/09540090902733871

Horgan, T., & Tienson, J. (1996). *Connectionism and the philosophy of psychology*. Cambridge, MA: MIT Press.

Houston, S. (2004). *Play piano in a flash!* New York: Hyperion.

Howell, P., Cross, I., & West, R. (1985). *Musical structure and cognition*. London, UK: Orlando Academic Press.

Hubel, D. H., & Wiesel, T. N. (1959). Receptive fields of single neurones in the cat's striate cortex. *Journal of Physiology, 148*, 574–591.

Hui, A. (2013). *The psychophysical ear: Musical experiments, experimental sounds, 1840–1910*. Cambridge, MA: MIT Press.

Humphrey, E. J., Bello, J. P., & LeCun, Y. (2013). Feature learning and deep architectures: New directions for music informatics. *Journal of Intelligent Information Systems, 41*(3), 461–481. doi: 10.1007/s10844-013-0248-5

Huron, D. (1999). *Music research using Humdrum: A user's guide*. Stanford, CA: Center for Computer Assisted Research in the Humanities.

Huron, D. B. (2006). *Sweet anticipation: Music and the psychology of expectation*. Cambridge, MA: MIT Press.

Isacoff, S. (2001). *Temperament: The idea that solved music's greatest riddle*. New York, NY: Alfred A. Knopf.

Isacoff, S. (2011). A natural history of the piano. New York, NY: Alfred A. Knopf.

Jun, S., Rho, S., & Hwang, E. (2010). Music retrieval and recommendation scheme based on varying mood sequences. *International Journal on Semantic Web and Information Systems*, *6*(2), 1–16. doi: 10.4018/jswis.2010040101

Katz, B. F. (1995). Harmonic resolution, neural resonance, and positive affect. *Music Perception*, *13*(1), 79–108.

Koelle, D. (2008). *The complete guide to JFugue: Programming music in Java*. www.jfugue.org.

Kohonen, T. (1977). *Associative memory: A system-theoretical approach*. New York, NY: Springer-Verlag.

Kohonen, T. (1984). *Self-organization and associative memory*. New York, NY: Springer-Verlag.

Kohonen, T. (2001). *Self-organizing maps* (3rd ed.). New York, NY: Springer.

Kohonen, T., Laine, P., Tiits, K., & Torkkola, K. (1991). A nonheuristic automatic composing method. In P. M. Todd & D. G. Loy (Eds.), *Music and connectionism* (pp. 229–242). Cambridge, MA: MIT Press.

Kosslyn, S. M. (1980). *Image and mind*. Cambridge, MA: Harvard University Press.

Kosslyn, S. M. (1994). *Image and brain*. Cambridge, MA: MIT Press.

Krumhansl, C. L. (1979). Psychological representation of musical pitch in a tonal context. *Cognitive Psychology*, *11*(3), 346–374. doi: 10.1016/0010-0285(79)90016-1

Krumhansl, C. L. (1990a). *Cognitive foundations of musical pitch*. New York, NY: Oxford University Press.

Krumhansl, C. L. (1990b). Tonal hierarchies and rare intervals in music cognition. *Music Perception*, 7(3), 309–324.

Krumhansl, C. L. (2005). The geometry of musical structure: A brief introduction and history. *ACM Computers In Entertainment*, *3*(4), 1–14.

Krumhansl, C. L., Bharucha, J. J., & Kessler, E .J. (1982). Perceived harmonic structure of chords in three related musical keys. *Journal of Experimental Psychology: Human Perception and Performance*, *8*(1), 24–36.

Krumhansl, C. L., & Kessler, E. J. (1982). Tracing the dynamic changes in perceived tonal organization in a spatial representation of musical keys. *Psychological Review*, *89*(4), 334–368.

Krumhansl, C. L., & Shepard, R. N. (1979). Quantification of the hierarchy of tonal functions within a diatonic context. *Journal of Experimental Psychology: Human Perception and Performance*, 5(4), 579–594.

Kruschke, J. K. (2011). *Doing Bayesian data analysis: A tutorial with R and BUGS*. Burlington, MA: Academic Press.

Laden, B., & Keefe, B. H. (1989). The representation of pitch in a neural net model of pitch classification. *Computer Music Journal, 13*, 12–26.

Laitz, S. G. (2008). *The complete musician: An integrated approach to tonal theory, analysis, and listening* (2nd ed.). New York, NY: Oxford University Press.

Large, E. W., & Kolen, J. F. (1994). Resonance and the perception of musical meter. *Connection Science, 6*, 177–208.

Larochelle, H., Mandel, M., Pascanu, R., & Bengio, Y. (2012). Learning algorithms for the classification restricted Boltzmann machine. *Journal of Machine Learning Research, 13*, 643–669.

Larson, W. S. (1930). Measurement of musical talent for the prediction of success in instrumental music. *Psychological Monographs, 40*(1), 33–73.

Leahey, T. H. (1987). *A history of psychology* (2nd ed.). Englewood Cliffs, NJ: Prentice-Hall.

Leighton, J. P., & Dawson, M. R. W. (2001). A parallel distributed processing model of Wason's selection task. *Cognitive Systems Research, 2*, 207–231.

Leman, M. (1991). The ontogenesis of tonal semantics: Results of a computer study. In P. M. Todd & D. G. Loy (Eds.), *Music and connectionism* (pp. 100–127). Cambridge, MA: MIT Press.

Lerdahl, F. (2001). *Tonal pitch space*. New York, NY: Oxford University Press.

Lerdahl, F., & Jackendoff, R. (1983). *A generative theory of tonal music*. Cambridge, MA: MIT Press.

Lettvin, J. Y., Maturana, H. R., McCulloch, W. S., & Pitts, W. H. (1959). What the frog's eye tells the frog's brain. *Proceedings of the IRE, 47*(11), 1940–1951.

Levine, M. (1989). *The jazz piano book*. Petaluma, CA: Sher Music Co.

Lewandowsky, S. (1993). The rewards and hazards of computer simulations. *Psychological Science, 4*, 236–243.

Lewin, D. (2007). *Generalized musical intervals and transformations*. New York, NY: Oxford University Press.

Lewis, J. P. (1991). Creation by refinement and the problem of algorithmic music composition. In P. M. Todd & D. G. Loy (Eds.), *Music and connectionism* (pp. 212–228). Cambridge, MA: MIT Press.

Lindsay, P. H., & Norman, D. A. (1972). *Human information processing*. New York, NY: Academic Press.

Lippmann, R. P. (1989, November). Pattern classification using neural networks. *IEEE Communications Magazine*, 47–64.

Liu, N. H., Hsieh, S. J., & Tsai, C. F. (2010). An intelligent music playlist generator based on the time parameter with artificial neural networks. *Expert Systems with Applications, 37*(4), 2815–2825. doi: 10.1016/j.eswa.2009.09.009

Longuet-Higgins, H. C., & Steedman, M. J. (1971). On interpreting Bach. In B. Meltzer & D. Michie (Eds.), *Machine intelligence* (Vol. 6, pp. 221–239). Edinburgh, UK: Edinburgh University Press.

Longyear, R. M. (1988). *Nineteenth-century Romanticism in music* (3rd ed.). Englewood Cliffs, N.J.: Prentice Hall.

Loy, D. G. (1991). Connectionism and musiconomy. In P. M. Todd & D. G. Loy (Eds.), *Music and connectionism* (pp. 20–36). Cambridge, MA: MIT Press.

Lunneborg, C. E. (1994). *Modeling experimental and observational data*. Belmont, CA: Duxbury Press.

Malmberg, C. F. (1918). The perception of consonance and dissonance. *Psychological Monographs*, *25*(2), 93–133.

Mammone, R. J. (1993). *Artificial neural networks for speech and vision*. New York, NY: Chapman & Hall.

Marolt, M. (2004a). A connectionist approach to automatic transcription of polyphonic piano music. *IEEE Transactions on Multimedia*, *6*(3), 439–449. doi: 10.1109/tmm.2004.827507

Marolt, M. (2004b). Networks of adaptive oscillators for partial tracking and transcription of music recordings. *Journal of New Music Research*, *33*(1), 49–59. doi: 10.1076/jnmr.33.1.49.35391

Marr, D. (1982). *Vision*. San Francisco, CA: W.H. Freeman.

Martineau, J. (2008). *The elements of music*. New York, NY: Walker & Company.

Marwala, T. (2000). Damage identification using committee of neural networks. *Journal of Engineering Mechanics-Asce*, *126*(1), 43–50. doi: 10.1061/(asce)0733-9399(2000)126:1(43)

McClelland, J. (1998). Connectionist models and Bayesian inference. In M. Oaksford & N. Chater (Eds.), *Rational models of cognition* (pp. 21–53). Oxford, UK: Oxford University Press.

McClelland, J. L., & Rumelhart, D. E. (1986). *Parallel distributed processing* (Vol. 2). Cambridge, MA: MIT Press.

McCloskey, M. (1991). Networks and theories: The place of connectionism in cognitive science. *Psychological Science*, *2*, 387–395.

McDermott, J., & Hauser, M. (2004). Are consonant intervals music to their ears? Spontaneous acoustic preferences in a nonhuman primate. *Cognition*, *94*(2), B11–B21. doi: 10.1016/j.cognition.2004.04.004

McLachlan, N., Marco, D., Light, M., & Wilson, S. A. (2013). Consonance and pitch. *Journal of Experimental Psychology-General*, *142*(4), 1142–1158. doi: 10.1037/a0030830

Medler, D. A., & Dawson, M. R. W. (1994). Training redundant artificial neural networks: Imposing biology on technology. *Psychological Research*, *57*, 54–62.

Medler, D. A., Dawson, M. R. W., & Kingstone, A. (2005). Functional localization and double dissociations: The relationship between internal structure and behavior. *Brain and Cognition*, *57*, 146–150.

Miller, G. A. (2003). The cognitive revolution: A historical perspective. *Trends in Cognitive Sciences*, *7*(3), 141–144.

Minsky, M. L., & Papert, S. (1969). *Perceptrons: An introduction to computational geometry*. Cambridge, MA: MIT Press.

Mohamed, A., Dahl, G. E., & Hinton, G. E. (2012). Acoustic modeling using deep belief networks. *IEEE Transactions on Audio Speech and Language Processing*, *20*(1), 14–22. doi: 10.1109/tasl.2011.2109382

Monterola, C., Abundo, C., Tugaff, J., & Venturina, L.E. (2009). Prediction of potential hit song and musical genre using artificial neural networks. *International Journal of Modern Physics C*, *20*(11), 1697–1718. doi: 10.1142/s0129183109014680

Moorhead, I. R., Haig, N. D., & Clement, R. A. (1989). An investigation of trained neural networks from a neurophysiological perspective. *Perception*, *18*, 793–803.

Mostafa, M. M., & Billor, N. (2009). Recognition of Western style musical genres using machine learning techniques. *Expert Systems with Applications*, *36*(8), 11378–11389. doi: 10.1016/j.eswa.2009.03.050

Mozer, M.C. (1991). Connectionist music composition based on melodic, stylistic, and psychophysical constraints. In P.M. Todd & D.G. Loy (Eds.), *Music and connectionism* (pp. 195–211). Cambridge, MA: MIT Press.

Mozer, M.C. (1994). Neural network music composition by prediction: Exploring the benefits of psychoacoustic constraints and multi-scale processing. *Connection Science*, *6*, 247–280.

Mozer, M.C., & Smolensky, P. (1989). Using relevance to reduce network size automatically. *Connection Science*, *1*, 3–16.

Munoz-Exposito, J. E., Garcia-Galan, S., Ruiz-Reyes, N., & Vera-Candeas, P. (2007). Adaptive network-based fuzzy inference system vs. other classification algorithms for warped LPC-based speech/music discrimination. *Engineering Applications of Artificial Intelligence*, *20*(6), 783–793. doi: 10.1016/j.engappai.2006.10.007

Nagashima, T., & Kawashima, J. (1997). Experimental study on arranging music by chaotic neural network. *International Journal of Intelligent Systems*, *12*(4), 323–339.

Neisser, U. (1967). *Cognitive psychology*. New York, NY: Appleton-Century-Crofts.

Newell, A. (1980). Physical symbol systems. *Cognitive Science*, *4*, 135–183.

Newell, A., & Simon, H. A. (1961). Computer simulation of human thinking. *Science*, *134*(349), 2011–2017.

Newell, A., & Simon, H. A. (1972). *Human problem solving*. Englewood Cliffs, NJ: Prentice-Hall.

Nickerson, C. M., Bloomfield, L. L., Dawson, M. R. W., Charrier, I., & Sturdy, C. B. (2007). Feature weighting in "chick-a-dee" call notes of Poecile atricapillus. *Journal of the Acoustical Society of America, 122*(4), 2451–2458. doi: 10.1121/1.2770540

Nisenson, E. (2000). *The making of* Kind of Blue*: Miles Davis and his masterpiece*. New York, NY: St. Martin's Press.

Norman, D. A. (1998). *The invisible computer*. Cambridge, MA: MIT Press.

Norman, D. A. (2002). *The design of everyday things*. New York: Basic Books.

Norman, D. A. (2004). *Emotional design: Why we love (or hate) everyday things*. New York, NY: Basic Books.

Oaksford, M., & Chater, N. (1991). Against logicist cognitive science. *Mind & Language, 6*, 1–38.

Oaksford, M., & Chater, N. (2007). *Bayesian rationality: The probabilistic approach to human reasoning*. New York, NY: Oxford University Press.

Omlin, C. W., & Giles, C. L. (1996). Extraction of rules from discrete-time recurrent neural networks. *Neural networks, 9*, 41–52.

Page, M. P. A. (1994). Modeling the perception of musical sequences with self-organizing neural networks. *Connection Science, 6*, 223–246.

Pao, Y. -H. (1989). *Adaptive pattern recognition and neural networks*. Reading, MA: Addison-Wesley.

Patel, A. D. (2008). *Music, language, and the brain*. New York, NY: Oxford University Press.

Pesic, P. (2010). Hearing the irrational: Music and the development of the modern concept of number. *Isis, 101*(3), 501–530.

Plantinga, J., & Trehub, S. E. (2014). Revisiting the innate preference for consonance. *Journal of Experimental Psychology-Human Perception and Performance, 40*(1), 40–49. doi: 10.1037/a0033471

Plantinga, L. (1984). *Romantic music: A history of musical style in nineteenth-century Europe*. New York, NY: W.W. Norton.

Plomp, R., & Levelt, W. J. M. (1965). Tonal consonance and critical bandwidth. *Journal of the Acoustical Society of America, 38*(4), 548–560.

Pollack, J. B. (1990). Recursive distributed representations. *Artificial Intelligence, 46*, 77–105.

Porter, L. (1998). *John Coltrane: His life and music*. Ann Arbor: University of Michigan Press.

Pylyshyn, Z. W. (1980). Computation and cognition: Issues in the foundations of cognitive science. *Behavioral and Brain sciences, 3*(1), 111–132.

Pylyshyn, Z. W. (1984). *Computation and cognition*. Cambridge, MA: MIT Press.

Ramsey, W., Stich, S. P., & Rumelhart, D. E. (1991). *Philosophy and connectionist theory*. Hillsdale, NJ: Lawrence Erlbaum Associates.

Restle, F. (1971). *Mathematical models in psychology: An introduction*. Baltimore, MD: Penguin.

Révész, G. (1954). *Introduction to the psychology of music*. Norman: University of Oklahoma Press.

Reynolds, A. G., & Flagg, P. W. (1977). *Cognitive psychology*. Cambridge, MA: Winthrop.

Ripley, B. D. (1996). *Pattern recognition and neural networks*. Cambridge, UK: Cambridge University Press.

Roig-Francolí, M. A. (2008). *Understanding post-tonal music*. Boston: McGraw-Hill.

Rojas, R. (1996). *Neural networks: A systematic exploration*. Berlin: Springer.

Rosch, E. (1975). Cognitive reference points. *Cognitive Psychology*, *7*(4), 532–547.

Rosch, E., & Mervis, C. B. (1975). Family resemblances: Studies in the internal structure of categories. *Cognitive Psychology*, *7*, 573–605.

Rosenblatt, F. (1958). The perceptron: A probabilistic model for information storage and organization in the brain. *Psychological Review*, *65*(6), 386–408.

Rosenblatt, F. (1962). *Principles of neurodynamics*. Washington, DC: Spartan Books.

Rowe, R. (2001). *Machine musicianship*. Cambridge, MA: MIT Press.

Rumelhart, D. E., Hinton, G. E., & Williams, R. J. (1986). Learning representations by back-propagating errors. *Nature*, *323*, 533–536.

Rumelhart, D. E., & McClelland, J. (1986a). On learning the past tenses of English verbs. In J. McClelland & D. E. Rumelhart (Eds.), *Parallel distributed processing. Volume 2: Psychological and biological models* (pp. 216–271). Cambridge, MA: MIT Press.

Rumelhart, D. E., & McClelland, J. (1986b). *Parallel distributed processing* (Vol. 1). Cambridge, MA: MIT Press.

Sano, H., & Jenkins, B. K. (1989). A neural network model for pitch perception. *Computer Music Journal*, *13*(3), 41–48. doi: 10.2307/3680010

Sapp, C. S. (2005). Visual hierarchical key analysis. *ACM Computers in Entertainment*, *3*(4), 1–19.

Sarikaya, R., Hinton, G. E., & Deoras, A. (2014). Application of deep belief networks for natural language understanding. *IEEE-ACM Transactions on Audio Speech and Language Processing*, *22*(4), 778–784. doi: 10.1109/taslp.2014.2303296

Scarborough, D. L., Miller, B. O., & Jones, J. A. (1989). Connectionist models for tonal analysis. *Computer Music Journal*, *13*(3), 49–55.

Schmajuk, N. A. (1997). *Animal learning and cognition: A neural network approach*. New York, NY: Cambridge University Press.

Schmuckler, M. A., & Tomovski, R. (2005). Perceptual tests of an algorithm for musical key-finding. *Journal of Experimental Psychology: Human Perception and Performance*, *31*(5), 1124–1149. doi: 10.1037/0096-1523.31.5.1124

Schoenberg, A. (1969). *Structural functions of harmony* (Rev. ed.). New York, NY: W. W. Norton.

Schumacher, M., Rossner, R., & Vach, W. (1996). Neural networks and logistic regression. *Computational Statistics & Data Analysis, 21*(6), 661–682. doi: 10.1016/0167-9473(95)00032-1

Schwab, E.C., & Nusbaum, H.C. (1986). *Pattern recognition by humans and machines: Visual perception* (Vol. 2). Orlando, FL: Academic Press.

Scimemi, B. (2002). The use of mechanical devices and numerical algorithms in the 18th century for the equal temperament of the musical scale. In G. Assayag, H. G. Feichtinger & J.-F. Rodrigues (Eds.), *Mathematics and music* (pp. 49–64). New York: Springer.

Searle, J. R. (1984). *Minds, brains and science*. Cambridge, MA: Harvard University Press.

Seashore, C. E. (1915). The measurement of musical talent. *The Musical Quarterly, 1*(1), 129–148.

Seashore, C. E. (1936). *Objective analysis of musical performance*. Iowa City, IA: The University Press.

Seashore, C. E. (1938/1967). *Psychology of music*. New York: Dover.

Seidenberg, M. (1993). Connectionist models and cognitive theory. *Psychological Science, 4*, 228–235.

Serafine, M. L. (1988). *Music as cognition: The development of thought in sound*. New York, NY: Columbia University Press.

Setiono, R., Baesens, B., & Mues, C. (2011). Rule extraction from minimal neural networks for credit card screening. *International Journal of Neural Systems, 21*(4), 265–276. doi: 10.1142/s0129065711002821

Setiono, R., Thong, J. Y. L., & Yap, C. S. (1998). Symbolic rule extraction from neural networks: An application to identifying organizations adopting IT. *Information & Management, 34*(2), 91–101. doi: 10.1016/s0378-7206(98)00048-2

Shanahan, D., & Broze, Y. (2012). A diachronic analysis of harmonic schemata in jazz. Paper presented at the 12th International Conference on Music Perception and Cognition, Thessaloniki, Greece.

Shapin, S. (1996). *The Scientific Revolution*. Chicago, IL: University of Chicago Press.

Shepard, R. N. (1964). Circularity in judgments of relative pitch. *Journal of the Acoustical Society of America, 36*(12), 2346–53. doi: 10.1121/1.1919362

Shepard, R. N. (1984). Ecological constraints on internal representation: Resonant kinematics of perceiving, imagining, thinking, and dreaming. *Psychological Review, 91*, 417–447.

Shepard, R. N. (1990). *Mind sights: Original visual illusions, ambiguities, and other anomalies*. New York, NY: W.H. Freeman and Co.

Shepherd, A. J. (1997). *Second-order methods for neural networks*. London, UK: Springer.

Shibata, N. (1991). A neural network-based method for chord note scale association with melodies. *Nec Research & Development, 32*(3), 453–459.

Shmulevich, I., & Yli-Harja, O. (2000). Localized key finding: Algorithms and applications. *Music Perception, 17*(4), 531–544.
Siegelmann, H. T. (1999). *Neural networks and analog computation: Beyond the Turing limit*. Boston, MA: Birkhauser.
Siegelmann, H. T., & Sontag, E. D. (1991). Turing computability with neural nets. *Applied Mathematics Letters, 4*, 77–80.
Simon, H. A. (1969). *The sciences of the artificial*. Cambridge, MA: MIT Press.
Sloboda, J. A. (1985). *The musical mind: The cognitive psychology of music*. Oxford, UK: Oxford University Press.
Smolensky, P. (1988). On the proper treatment of connectionism. *Behavioral and Brain Sciences, 11*, 1–74.
Smolensky, P., & Legendre, G. (2006). *The harmonic mind: From neural computation to optimality-theoretic grammar*. Cambridge, MA: MIT Press.
Snyder, B. (2000). *Music and memory: An introduction*. Cambridge, MA: MIT Press.
Sperry, R. W. (1993). The impact and promise of the cognitive revolution. *American Psychologist, 48*(8), 878–885.
Stanton, H. M., & Seashore, C. E. (1935). *Measurement of musical talent: The Eastman experiment*. Iowa City, IA: The University Press.
Stephenson, B. (1994). *The music of the heavens: Kepler's harmonic astronomy*. Princeton, NJ: Princeton University Press.
Stevens, C., & Latimer, C. (1992). A comparison of connectionist models of music recognition and human performance. *Minds and Machines, 2*(4), 379–400.
Straus, J. N. (1991). A primer for atonal set theory. *College Music Symposium, 31*(1–26).
Straus, J. N. (2005). *Introduction to post-tonal theory* (3rd ed.). Upper Saddle River, NJ: Prentice Hall.
Sudnow, D. (1978). *Ways of the hand: The organization of improvised conduct*. Cambridge, MA: Harvard University Press.
Sullivan, J. W. N. (1927). *Beethoven: His spiritual development*. London, UK: J. Cape.
Taha, I. A., & Ghosh, J. (1999). Symbolic interpretation of artificial neural networks. *IEEE Transactions on Knowledge and Data Engineering, 11*(3), 448–463. doi: 10.1109/69.774103
Takane, Y., Oshima-Takane, Y., & Shultz, T. R. (1994). Approximations of nonlinear functions by feed-forward neural networks Proceedings of the Japan Classification Society Meeting (pp. 26–33). Tokyo: The Japan Classification Society.
Temperley, D. (1999). What's key for key? The Krumhansl-Schmuckler key-finding algorithm reconsidered. *Music Perception, 17*(1), 65–100.
Temperley, D. (2001). *The cognition of basic musical structures*. Cambridge, MA: MIT Press.
Temperley, D. (2004). Bayesian models of musical structure and cognition. *Musicae Scientiae, 8*(2), 175–205.

Temperley, D. (2007). *Music and probability*. Cambridge, MA: MIT Press.
Temperley, D., & Marvin, E. W. (2008). Pitch-class distribution and the identification of key. *Music Perception, 25*(3), 193–212. doi: 10.1525/mp.2008.25.3.193
Terhardt, E., Stoll, G., & Seewann, M. (1982a). Algorithm for extraction of pitch and pitch salience from complex tonal signals. *Journal of the Acoustical Society of America, 71*(3), 679–688.
Terhardt, E., Stoll, G., & Seewann, M. (1982b). Pitch of complex signals according to virtual-pitch theory: Tests, examples, and predictions. *Journal of the Acoustical Society of America, 71*(3), 671–678.
Thomas, J. C. (1975). *Chasin' the trane: The music and mystique of John Coltrane* (1st ed.). Garden City, N.Y.: Doubleday.
Thrun, S. (1995). Extracting rules from artificial neural networks with distributed representations. In G. Tesauro, D. S. Touretzky & T. K. Leen (Eds.), *Advances in neural information processing systems* (Vol. 7, pp. 505–512). Cambridge, MA: MIT Press.
Todd, P. M. (1989). A connectionist approach to algorithmic composition. *Computer Music Journal, 13*(4), 27–43.
Todd, P. M., & Loy, D. G. (1991). *Music and connectionism*. Cambridge, MA: MIT Press.
Tymoczko, D. (2006). The geometry of musical chords. *Science, 313*(5783), 72–74.
Tymoczko, D. (2008). Scale theory, serial theory and voice leading. *Music Analysis, 27*(1), 1–49. doi: 10.1111/j.1468-2249.2008.00257.x
Tymoczko, D. (2011). *A geometry of music: Harmony and counterpoint in the extended common practice* (E-pub ed.). New York, NY: Oxford University Press.
Tymoczko, D. (2012). The generalized Tonnetz. *Journal of Music Theory, 56*(1), 1–52. doi: 10.1215/00222909-1546958
Van Egmond, R., & Butler, D. (1997). Diatonic connotations of pitch-class sets. *Music Perception, 15*(1), 1–29.
Van Gelder, T. (1991). What is the "D" in "PDP"? A survey of the concept of distribution. In W. Ramsey, S. P. Stich & D. E. Rumelhart (Eds.), *Philosophy and connectionist theory* (pp. 33–59). Hillsdale, NJ: Lawrence Erlbaum Associates.
Vos, P. G., & Troost, J. M. (1989). Ascending and descending melodic intervals: Statistical findings and their perceptual relevance. *Music Perception, 6*(4), 383–396.
Vos, P. G., & Van Geenen, E. W. (1996). A parallel-processing key-finding model. *Music Perception, 14*(2), 185–223.
Watson, J. B. (1913). Psychology as the behaviorist views it. *Psychological Review, 20*, 158–177.
Waugh, N. C., & Norman, D. A. (1965). Primary memory. *Psychological Review, 72*, 89–104.
Wechsler, H. (1992). *Neural networks for perception: Computation, learning, and architectures* (Vol. 2). Boston, MA: Academic Press.

Whittall, A. (1987). *Romantic music: A concise history from Schubert to Sibelius*. London, UK: Thames and Hudson.

Wieczorkowska, A. A., & Kubera, E. (2010). Identification of a dominating instrument in polytimbral same-pitch mixes using SVM classifiers with non-linear kernel. *Journal of Intelligent Information Systems, 34*(3), 275–303. doi: 10.1007/s10844-009-0098-3

Winston, P. H. (1975). *The psychology of computer vision*. New York, NY: McGraw-Hill.

Wood, G. (2002). *Living dolls: A magical history of the quest for artificial life*. London, UK: Faber and Faber.

Yaremchuk, V., & Dawson, M. R. W. (2005). Chord classifications by artificial neural networks revisited: Internal representations of circles of major thirds and minor thirds. Artificial Neural Networks: Biological Inspirations – Icann 2005, Pt. 1, Proceedings, 3696, 605–610.

Yaremchuk, V., & Dawson, M. R. W. (2008). Artificial neural networks that classify musical chords. *International Journal of Cognitive Informatics and Natural Intelligence*, 2(3), 22–30.

Zhao, Z. Q., Huang, D. S., & Sun, B. Y. (2004). Human face recognition based on multi-features using neural networks committee. *Pattern Recognition Letters*, 25(12), 1351–1358. doi: 10.1016/j.patrec.2004.05.008

Index